3rd Edition

Tune In On Telephone Calls!

by Tom Kneitel, K2AES

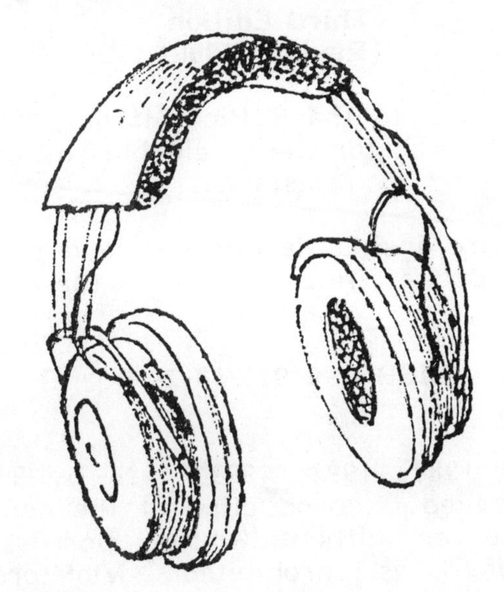

CRB Research Books, Inc.

P.O. Box 56, Commack, New York 11725

Dedicated to:
Terri E. Kneitel, M.S., P.E.,
my favorite engineer.

Cover designed by:
Robin L. Smith,
Art Director,
CRB Research Books, Inc.

Third Edition
(Revised 1996)

The author wishes to thank Rick Maslau, KNY2GL, and Harry Caul, KIL9XL, for their valuable cooperation and assistance.

Cover and entire inside design and layout prepared by CRB Research Books, Inc., Printed in the United States of America.

ISBN: 0-939780-24-0

CONTENTS

1 1 1 1 1 1 1 1 1 1 1 1 1 1 1 1 1

Tuning In On Telephones

Ever since they began installing party line telephones in people's homes, in the early 1900's, folks would pass the time of day by a little harmless eavesdropping on their neighbors' telephone calls. Listening in on the party line-- and turning out the lights and sitting by an open window while the next door neighbors were having a particularly juicy row-- became deeply ingrained in our way of life. Sure, maybe a little sneaky-- like driving 5 m.p.h. above the posted speed limit, or holding a misdirected letter up to the light to see what it's all about, or peeking in someone else's medicine cabinet while nobody's looking, but not of any real negative consequence.

It may or may not be coincidence that in the late 1940's, when mobile radio in the 35 and 152 MHz bands first became available to the public, the first low cost 30 to 50 MHz and 152 to 174 MHz tunable radios also went on sale. These were a great new replacement for the old party line trick, and came along just when party lines were beginning to diminish in popularity. Of course, to those who owned shortwave receivers, none of this was new-- 2 MHz ship-to-shore telephone calls, as well as higher frequency transoceanic and high seas radio phone conversations had been easily available for their listening for a very long time. Something to satisfy the little bit of peepingtomism most folks have-- the conversations were there anyway, so why not check them out? Even if nobody listened, they'd still be there. Wouldn't they?

I didn't put them there. You didn't put them there. If people didn't want others to know what they were talking about, then maybe they should know better than yak about those things over a radio transmitter-- or get a voice scrambler to secure some privacy for themselves.

Interestingly, although international telephone calls,

An early (1946) single channel car phone for the 152 MHz band was found to be a more inexpensive way for small taxi companies to dispatch than setting up a private two-way system of their own!

ship-to-shore calls, and calls from car phones have been on the scene for decades, there really never was any big, serious flap about communications privacy or security. Everybody knew that others could be-- would be-- listening in on their conversations and that went with the territory. Those who had private things to say either let everybody else in on their chatter, or else kept their traps shut, or figured out a more secure way of communicating.

For the government's interest, there had long been a law on the books known as Section 605 (now called Section 705) of the Communications Act. Basically, it specified that if a person overheard someone else's radio communications there

Modern telecommunications technology causes many telephone calls to utilize radio waves at some point along the circuit.

was no real harm done, but the eavesdropper wasn't permitted to reveal to others or use for gain any information that came through. So, a person could listen all they wanted for their own personal interest, amusement, information, or hobby. And those who demaded privacy found that securing that status was their problem-- the airwaves are a natural resource and belong to the public.

You go to the beach and want to change, it's your problem to do it in a cabana if you want privacy. If you elect to change out in the open instead of using a cabana, it's hardly logical to think that you can ask-- or demand-- that everybody else on the beach divert their gaze away from you so that you can change in privacy. If anything, your loud demands for such privacy would probably attract a crowd to see what was going on and what all the fuss was about.

If you don't believe me, go to a public beach and yell out that nobody's allowed to watch you while you change because you want privacy. If that doesn't attract the attention of every person within earshot, nothing will. It's a guarantee.

Then comes along one more version of an old stunt. In the

For all of their deluxe features, the nitty gritty is that cellular car phones are two way radios that send out signals on the airwaves-- same as hams, fire departments, or taxi services. All can be heard by anybody with a scanner.

early 1980's, cellular mobile telephones (CMT's) arrived on the scene. Despite their technological peculiarities, CMT's are still your basic car phones, operating now in the 800 MHz band instead of on bands used previously. Aside from some fancy luxury features, they're still two-way radios that might be monitored by any person having a receiver or scanner that tuned the 800 MHz band. Had they just left well enough alone, there would have been a few busybodies and hobbyists tuned in on 800 MHz-- same crowd that has listened in on other car phones for years-- and there'd be no fuss.

But no, that wasn't good enough. The CMT industry, with mucho bucks, formed itself into a lobbying group and decided to go out on the beach and demand that everybody look the other way because they wanted privacy. At first they tried to get a law passed in California that said people weren't allowed to listen to the frequencies used by CMT's. California laughed them right out of the state.

Next, the CMT lobbyists showed up in Washington with their scheme, only now it had been embellished with all sorts of window dressing to make it appear that new legislation was needed to protect government snooping on private citizens, and other irrelevant poop.

Let's face it, any fat cat industry lobbyist can show up in Washington and with enough bluster and bull, can find at least a couple of gullible dummys to stand up and do their bidding. In this case, they had no trouble. Soon after, the "Electronic Communications Privacy Act of 1985" (H.R. 3378) was

sponsored by Rep. Robert Kastenmeier (D-Wisconsin), and Rep. Carlos Moorhead (R-California). A bill with the same name and almost identical wording (S-1667) was presented to the Senate by Sen. Pat Leahy (D-Vermont).

Rep. Kastenmeier seemed especially revved up on the whole federal snooping smokescreen of the legislation. The CMT played on him like he was a finely tuned antique violin. Although the legislation was thought up and pushed only by the CMT industry, Kastenmeier took the entire bait with gusto, making preposterous pronouncements such as, "Today we have large-scale electronic mail operations, cellular and cordless telephones, paging devices, miniaturized transmitters for radio

In 1929, U.S. Secretary of State Henry Stimson complained about government's espionage activities, observing, "Gentlemen do not read each other's mail." Nevertheless, it is considered honorable by our government today. The NSA, CIA, and other federal agencies have ground stations and secret satellites (variously known by code names such as Magnum, Aquacade, Argus, Rhyolite, Vortex, Chalet, etc.) that monitor the world's (including certain U.S.) telephone calls using radio circuits. Yes, despite even the ECPA!

surveillance, lightweight compact television cameras for video surveillance, and a dazzling array of digitized information networks, which were little more than concepts two decades ago...This array of technologies enhances the risk that our communications will be intercepted by either private parties or the government."

It was the tired old trick of using a supposedly and apparently well-intentioned noble motive as a red herring to draw applause while something basically insidious was being sneaked through hidden in the herring's gullet. Those who might spot the true core of the proposed legislation are supposed to be afraid to rip it for fear that they'll be accused of endorsing federal agency snooping on private citizens. This trick works, like hiding the aspirin in baby's orange juice.

When stripped of its padding about cutting federal

surveillance, all the fuss was just so that the CMT industry could advise its customers that it didn't make any difference that their CMT transmissions were in the clear (unscrambled) since federal legislation was in effect that guaranteed privacy. To most members of the general public-- people who believe when they're told 8 cylinder cars will offer 45 m.p.g., and Munchy Fiber cereal tastes good and will make a 60 year old guy feel like 20 again-- the promise of privacy by federal declaration would sound logical.

Can you imagine the dude on the beach getting ready to change, demanding that others look away because he was waving around a piece of paper upon which was written a law that said nobody should watch him?

Well, the gist of the originally proposed legislation mostly said that it would be a violation of the Electronic Communications Privacy Act (ECPA) for people to listen to most radio transmissions except broadcasting, ham, and CB stations.

Those who owned scanner and shortwave receivers were both scared and outraged that such a law would or could be seriously considered-- why the CMT industry wasn't sent packing with their scuzzy law and told that if they wanted to assure privacy, then let their service use voice scramblers-- that communications security for their subscribers was their own responsibility, not that of casual listeners who had freely monitored the public's airways for more than eighty years.

To be sure, many letters and petitions decrying the proposed ECPA were written to Washington. Numerous magazines cried out about the ECPA and its threat to the rights of the public at large. Not only that, but it was pointed out that there wasn't any way to detect violations of the law, or obtain evidence that the law was violated.

Moreover, even federal enforcement agencies appeared little interested in even attempting to bother with the law. Reporter David W. MacDougall, in The News and Courier/The Evening Post (a Charleston, SC newspaper) quoted a spokesman for the FBI office in Columbia, SC as stating that electronic eavesdropping of any kind was illegal, adding, "With these cellular phones, it would be real difficult to prosecute. If somebody were doing it on a regular basis, or if someone was being paid to listen to phone conversations, we would want to go after them."

Bob Grove, Editor of Monitoring Times, was quoted as stating, "It's completely unenforceable. The FBI is the agency in charge of enforcement and they have gone on record saying they will not enforce it except in the most egregious circumstances, such as blackmail."

Despite the outcry, the letters, the hearings that were held, it seemed that most of the people in Washington had no real concept of what the ECPA was all about. Or possibly they realized that it was no more than one more piece of meaningless "junk" legislation that, if nothing else, would serve to mollify some industry's high-pressure lobbyists in the hopes that maybe they'd pack up and finally leave Washington.

For whatever reasons, the ECPA, in a considerably toned down but still absurd version, was 1-2-3 railroaded through both houses of Congress and rubber stamped into law by the 99th Congress in the final hours that body was in session-- in fact the session ran two weeks late and the festive legislators were walking around wearing buttons reading, "Free the 99th Congress."

They were cleaning out the cupboards and passing everything into law just to close up and go home. So desperate were they to finish up that they would have voted a dead cow into law. At one point the ECPA was even incorporated into pending drug control legislation. It was eventually reinstituted as an independent piece of legislation where it was offered to the U.S. Senate Judiciary Committee for their consideration. After spending exactly 25 seconds considering its intent, merits, and ramifications, the bill was unanimously approved and went from there through its final approval by both Houses of Congress faster than a nudist with hot soup in his lap.

You really can't help but notice the sleaziness of everything surrounding the ECPA and everybody connected with this tatty little piece of work. It's not that you don't understand what some of these politicians are doing, it's that you fear that they don't either!

And Rep. Kastenmeier was still cranking out hot air, telling one of his constituents who wrote to complain about the ECPA (Terry O'Laughlin, ham WB9GVB), "The bill is designed to extend the protection of the Wiretap Act...to new modes of communication, such as computer transmissions by satellite, as well as cellular telephones."

There is no record as to how Kastenmeier was able to deal

PRIVACY: THE REALITY

The Court of Appeals for the Third Circuit decided a case involving a car phone. A man was using his car phone to talk to his lawyer--he was discussing details of criminal activity. A scanner owner was eavesdropping and taped the chat, then gave the tapes to the U.S. Attorney. The lawyer and his client sued the scanner owner asking damages for federal law violations. The point was whether the plaintiffs "possessed a subjective expectation of privacy that was also objectively reasonable." Plaintiffs said that regardless of the fact that a car phone can be easily monitored on a scanner, people using car phones do have an expectation of privacy and don't think that someone will be taping what's being said and turning the tapes over to the U.S. Attorney.

The court dismissed the case, siding with the scanner owner, saying that people who are talking over the airwaves can't reasonably expect privacy. (Edwards v. State Farm Insurance Co., CA 5, (Garwood, J.), No. 86-3686, 12/7/87, 56 LW 2345, 12/22/87).

The Calif. Public Utilities Commission is also unconvinced about car phone privacy. It asked car phone services to tell their subscribers to use scramblers if they need privacy since calls may not be completely private. Car phone users were also told to advise the person to whom they're speaking about "the privacy issue at the beginning of each conversation."

with the public's free access to car phone communications for the previous 40 years, or ship-to-shore, high seas, and transoceanic telephone calls that had existed since the 1930's and earlier.

The way the final version of the ECPA (Public Law 99-508 of October 21, 1986) looked, it is illegal to monitor voice paging systems, SCA subcarriers on FM broadcast stations, any communications that used coded or scrambled or other techniques deliberately employed to assure privacy, remote broadcast or studio-transmitter links, private microwave transmission systems, or Common Carriers-- a Common Carrier being a communications service available to the general public for hire, such as car phones of all kinds.

In the area of telephone calls, the ECPA doesn't bear upon cordless telephones, tone-only paging signals, and systems relating to aircraft and maritime operations.

There is still no way of detecting violations of the ECPA, no way I can think of to obtain real evidence of violations, and still there have been no indications that any federal agency is even slightly inclined to bother with attempting to enforce the thing. If anything, federal agencies have stated flat out that they have neither the time, resources, nor manpower to devote to routine violations of the ECPA.

Equipment that receives all Common Carrier frequencies, including CMT's, is in the public's hands, and easily obtainable from many sources. Moreover, monitoring CMT's and other Common Carriers has, of late, enjoyed a rapidly growing band of diehard, and even fanatic enthusiasts, plus many casual listeners who mix in shortwave, VHF and UHF telephone call monitoring in with their other general receiving efforts. And, of course, there are amateur and professional monitors who listen with motives of their own.

The ECPA tars all of these people with the same brush.

This book is a listing of all of the many channels currently being used for telephone calls and paging in North America, on all bands. You may not be aware of some of these frequencies and services, and undoubtedly new stations, frequencies, and services will be heading down the street as the communications explosion continues to permeate our society-- such as the Basic Exchange Telecommunications Radio Service, which proposes to link some 450,000 rural American homes into the regular telephone systems by means of radio.

This service is expected to primarily benefit those in desert and mountain areas of the west, also some people in Appalachia. It is expected that 100+ telephone companies will jump on to the BETRS bandwagon, which will be opening up on frequencies not previously available for such activities.

In the book you'll come across several terms that might need some explanation at the outset. Like, "landline." Landline telephones are regular home and office telephones, connected to telco offices by overhead or underground copper wires or fiberoptic lines. A "patch" or "phone patch" is a telephone call that is received by radio at a two-way communications base station and then, via the facilities of that base station, fed out into landlines to an individual subscriber some distance away.

Note that systems, such as are listed herein, are in a constant state of change. Undoubtedly, with such a large amount of information, plus new frequencies, services, etc., you may well discover new information. I hope that you will pass along all such data to me in care of the publisher of this book.

Also be aware, that there are a number of private and non-public communications systems operating that have the ability to run mobile phone patches. While I have listed railroad PBX systems, which fit into this category, I have generally not attempted to list such units. However, there are federal agencies, for instance, like the FCC on 167.05 MHz, and the FAA on 166.175 MHz, that are equipped for patches. Also, in the area of Washington, DC there have been private VIP car phone calls reported on 172.365, 172.395, and 172.425 MHz. Companies, too, may have such communications facilities, and perhaps if sufficient information on these is received from readers, the information will be included in future editions of this book.

Remember when reading this book, that Section 705A of the Communications Act applies to all transmissions (except those on the ham bands), and the ECPA applies to car phones, voice pagers, and several other things. Good idea to obtain copies of these laws-- read them, understand them, and (of, course) observe them.

2 2 2 2 2 2 2 2 2 2 2 2 2 2 2 2 2

The Hardware

This isn't intended to be a book on the art and science of general use of a scanner or communications receiver. There is material available in books that is intended to provide the beginner with information on these topics. Also, periodicals such as **Popular Communications Magazine** and **Monitoring Times** regularly present colums and features on these topics, also showing advertisements for the latest scanners and shortwave communications receivers.

In addition, local communications equipment stores as well as mail order electronics suppliers are usually more than happy to answer all of your questions and help you to select the best equipment for the right price.

If you are a rank novice and don't know the difference between a scanner and a scandal (don't fret, we all started out that way), perhaps a little basic groundwork is in order so that you'll know what to shop for.

A communications receiver is the equipment used for receiving frequencies between 1,600 and 30,000 kiloHertz

Kenwood's R-2000 communications receiver is ideal for general monitoring of SSB-mode shortwave activities.

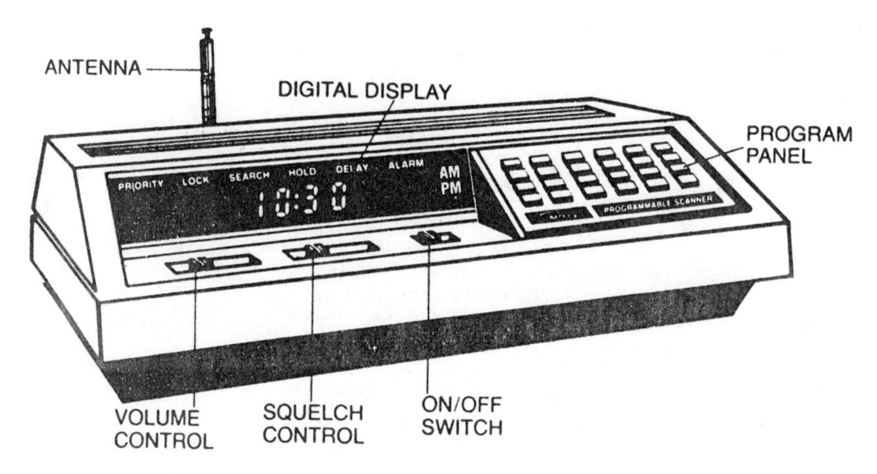

Basically, modern scanners are no more complicated than tape decks or television sets.

(kHz). This frequency range may also be referred to as 1.6 to 30 MegaHertz (MHz). Although there are many brands of equipment covering shortwave bands, not all of the sets are "general coverage," that is, can receive all shortwave frequencies. Some will receive only certain international shortwave broadcasting bands-- such sets will not receive any of the frequencies or bands shown in this book.

You'll want to be certain that the set you select is capable of receiving single sideband (SSB) mode signals. Also, get a set that has digital frequency display (usually by LED's or LCD's), otherwise you'll never be able to tune in a desired frequency with sufficient accuracy to find what you're looking for.

While the smaller transistor portable shortwave receivers are probably OK for listening to the BBC and other shortwave broadcasters, they aren't very good for locating and picking up shortwave SSB-mode two-way communications. Best bets are general coverage communications receivers made by ICOM, Kenwood, Yaesu, and (more expensively) Japan Radio Corp.

For an antenna, you may not need more than 50 to 100 feet of stranded, insulated, copper wire tossed out of the window. If you feel you need anything more formidable, check with dealers who sell commercial shortwave antennas.

A scanner is a receiver capable of picking up two-way FM communications. The frequency bands covered by most scanners are 30 to 50 MHz (known as the VHF low band); 150 to 174 MHz (VHF high band); 450 to 470 MHz (UHF band); and

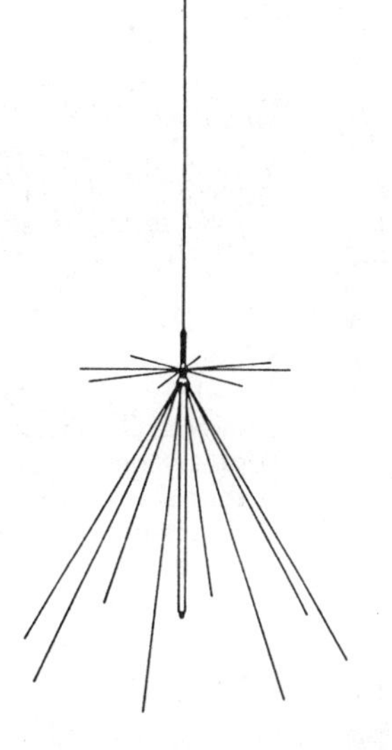

A discone is a scanner antenna that is designed to cover a wide range of frequencies, such as 25 through 1,300 MHz. Several companies now offer these. The one shown here is R.F. Limited's Palomar D-130 Super Wideband Discone. It's made of stainless steel.

470 to 512 MHz (UHF "T" band). Some scanners pick up additional bands and frequencies, such as 118 to 136 MHz (VHF aero band). A few newer models can receive frequencies as high as 1300 MHz, and that includes the cellular channels in the 800 to 900 MHz band-- we'll get to those in a minute.

Scanners are pretty simple to operate, and the instruction manuals packed in with each are written so that every aspect of the equipment's potentials is thoroughly explained in non-technical language. You can't go wrong. Just be certain that the unit you purchase will pick up the band you want-- that is to say, if you want to receive communications in the 800 to 900 MHz range, remember that not every scanner on the market can receive those frequencies.

Popular brands of scanners include Uniden Bearcat, Cobra, Regency, Fanon Courier, J.I.L., ACE, ICOM, Yaesu, Fox, and Radio Shack's Realistic brand.

Scanner antennas, for maximum reception, should be mounted outside and as high above the ground as possible. Safety considerations call for the mounting location to be sufficiently far from electric wires so that the antenna system cannot come into contact with the wires during the mounting process, or at some time after it is mounted. Electric wires can constitute a severe shock hazard if they should come into contact with any antenna system. Best bet is also to protect

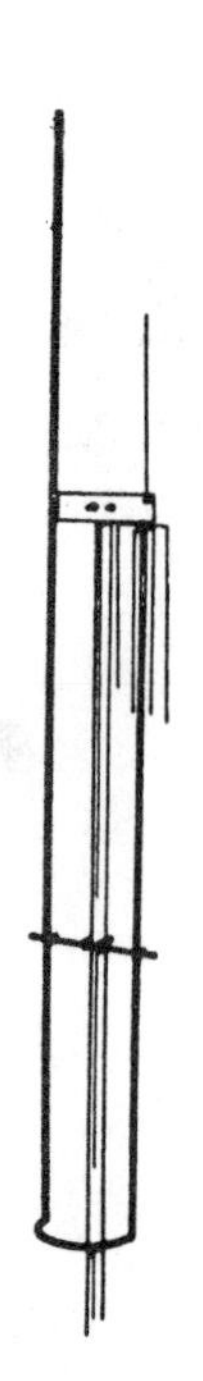

Two versatile scanner antennas. On the left is an Avanti AV-801 Astro Scan. It offers excellent results on communications bands from 25 through 512 MHz. It weighs 2 lbs. At the right is the Grove ANT-1B beam, a directional type antenna for long range scanner reception (in the direction the antenna is pointed) on all bands between 25 and 960 MHz. Can be used with a TV antenna rotor to change its direction of reception.

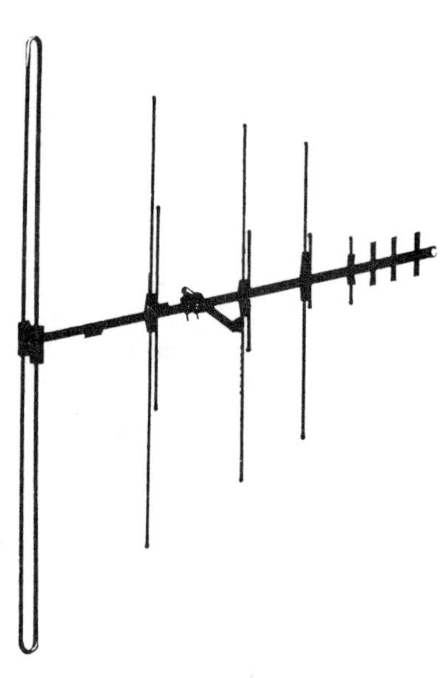

your scanner against lightning strikes-- a simple and inexpensive gizmo does the job, ask any communications dealer about lightning protection.

Many scanner antennas are designed so that they can receive signals over a sufficiently wide swath of signals (for instance 30 to 512 MHz) so that you'll need only one antenna to pick up virtually all scanner frequencies. Some newer models even include the 800 to 900 MHz band in this coverage. Popular scanner antennas are made by Antenna Specialists, Valor, Radio Shack, Enscomm, Grove, R.F. Limited Palomar, Butternut, and American Electronics. There are also several types of "indoor active" electronic antennas that can be used by those who are unable to put up an outside antenna. Of course, all scanners come with "built-in" antennas that (at the very least) should be suitable for local reception.

Generally speaking, when shopping for communications equipment, deal with communications shops or with mail order firms that specialize in communications equipment. You'll get better service, better prices, a wider selection of equipment, and more answers to your questions than you'll find at a general merchandise or so-called discount supplier selling

Uniden Bearcat's BC-800XLT scanner picks up all communications bands between 29 and 912 MHz, and that includes cellular phone channels.

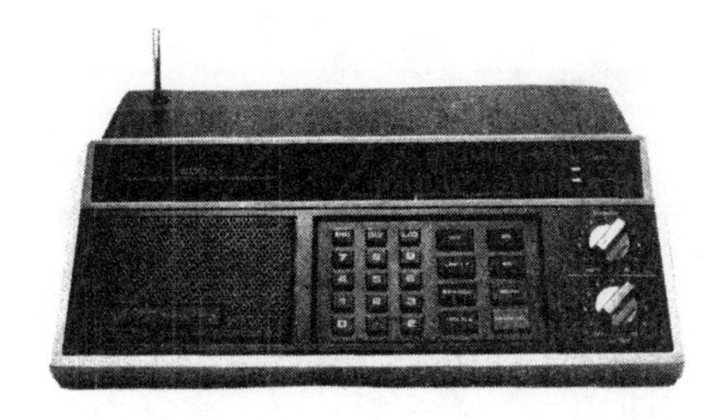

everything from hair dryers to lawn furniture and bracelets.

All scanners do not cover the 800 to 900 MHz band, where the cellphone frequencies are. Some recently produced models that do include the Radio Shack PRO's-24, 25, 26, 29, 34, 37 39, 43, 46, 51, 60, 62, 2004, 2005, 2006, 2022, 2026, 2027, 2030, 2032, 2035, 2037, 2038, 2039, 2040, 2042; Regency TS-2, MX-4000, HX-2200, R-4030; Trident TR's-980, 1200, 1200; AOR AR's 1500, 2800, 3000, 2500, 950/900, 1000, 2002; Yaesu FRG-9600; ICOM IC's-R1, R100, R7000, R7100; Uniden Bearcat BC's-800, 855, 890, 200/205, 760/950, 700A, 2500, 8500, MR-8100A. Some models had cellular bands blocked at the factory. In many older sets it is possible for owners to restore them using simple modifications.

The standard reference guides to user restoration of many popular scanners are the **Scanner Modification Handbooks**, published by CRB Research Books, Inc.

Converter accessories enabling reception of 800 to 900 MHz band frequencies on scanners that do not cover this band were readily available until recently. Sadly, these outboard devices are now history. Why?

It's because FCC regulations effective as of April, 1994, ended the manufacture of scanners capable of receiving, or being easily user-restored to receive, cellular bands. These regs also prohibited the manufacture of converters that can pick up cellular frequencies.

The Cellular Security Group is a professional technical service that can modify the circuitry of some

ADDENDUM

PRO-2004
PROGRAMMABLE SCANNER
General Coverage AM/FM Monitor Receiver
Cat No. 20-119

Dear Customer,

The unit is changed so the following frequencies are not received. When you try to enter the frequency in these ranges, ERROR will be displayed. The search function also skips these frequencies.

825.000 to 844.995 MHz
870.000 to 889.995 MHz

Radio Shack
Fort Worth, TX 76102

Printed in Japan
86D-6887

Radio Shack's excellent PRO-2004 scanner comes packed with this note to say that its ability to receive cellular frequencies has been removed. It's easy easy enough to restore that ability, however.

newer model blocked scanners. This results in what is essentially a switchable internal converter that will permit reception of signals over the entire 800 to 900 MHz spectrum, without gaps. Reception takes place in the scanner's 400 to 500 MHz band. Costs about $100. Check with them to see if your scanner can be done, and to get further information. Their phone number is: (508) 281-8892. Tell them I sent you.

There are several excellent scanner antennas that are great on these frequencies. You'll want to consider CRB Research's high-efficiency MAX-HH. This is a popular 800 to 900 MHz type for handheld scanners. For base station rooftop mounting, it's the MAX-CMP type. These are intended for serious eavesdropping, and can pull in weaker signals.

Attempting to monitor on the 800 MHz band with an antenna system that was not designed to receive those frequencies will result in reduced reception range. Signals on 800 MHz grow weak very quickly if they need to travel more than 25 ft. through RG-58/U coaxial

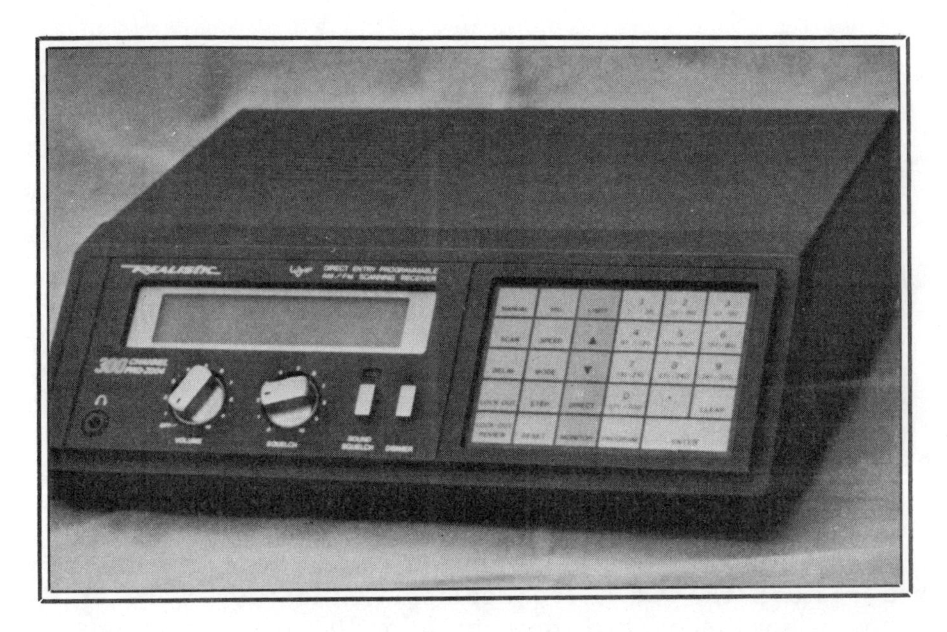

The Radio Shack Realistic PRO-2004 scanner.

cable, which may be what you are now using at your base station.

Keep cable lengths as short as possible. Use 52 ohm cable, preferably some kind of low-loss type intended for UHF use. If you can't easily get that, then at least use RG-8/U cable instead of the skinny stuff.

Some historic information is relevant at this point. Many scanner users regard the Radio Shack Realistic PRO-2004/2005/2006 series of scanners as being extremely versatile. They tune from 25 to 1300 MHz, have user selectable AM/NFM/WFM reception on all frequencies, quick scanning rates, lots of memory channels, and (in search/scan mode) the user can select from different search steps between 5 and 50 kHz. No longer made, they remain in wide use.

When the first model in this series, the PRO-2004, was originally announced (which was before the actual sets were shipped to local stores), the stats claimed that it would receive the bands used for CMT calls. But just about simultaneously, the ECPA was passed into law.

Radio Shack is in the CMT business, and the company supported the passage of the ECPA. With the ECPA passed, Radio Shack apparently had second thoughts about the ramifications.

How would it look to sell CMT's and also the scanners that could illegally eavesdrop on those frequencies? There was nothing illegal about making or selling such sets, so the decision was strictly voluntary. With the sets already built and in cartons, the only way to change the situation was to open each and every carton and remove every PRO-2004. Then, have a technician make a modification to each scanner that would knock out its ability to receive the CMT frequencies. This is exactly what was done, although what they did was such that it could easily be user-reversed.

It didn't take scanner owners long to figure out how to restore the CMT frequencies. While they were at it, they also discovered how to add 100 more channels to the memory capabilities of the PRO-2004, plus numerous additional performance enhancements.

As the PRO-2004 was eventually replaced by the PRO-2005, then the PRO-2006 models, scanner owners continued to modify these sets for CMT reception, and also added many other features. It also turned out that quite a few other popular scanners could also be relatively easily modified by scanner owners for CMT frequency restoration, as well as the addition or improvement of many other operating features.

3 3 3 3 3 3 3 3 3 3 3 3 3 3 3 3 3

Cellular Telephone Calls

There's no debating about the popularity of cellular mobile telephones (CMT's). Since the time the CMT service went into commercial operation in the early 1980's, it has captured the fancy of the communications-hungry public.

CMT's aren't only a useful communications tool, they're a great status symbol. Open an attache case and extract a CMT and you impress everyone in the room. Drive up in your BMW, Porsche, Corvette, Trans-Am, or Mercedes, and if the car didn't catch their eye, a CMT antenna on the vehicle's roof will certainly do the job-- two CMT antennas makes an even louder shout about your clout, importance, and/or wealth.

For those who'd like to fantasize about having that lifestyle, there are even fake CMT mobile antennas, dummy hollow plastic cellular phone lookalikes, and even CB transceivers and antennas that are reasonably good doubles for

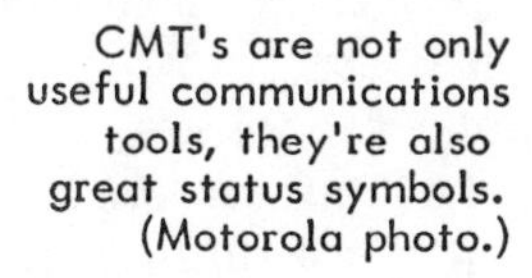

CMT's are not only useful communications tools, they're also great status symbols. (Motorola photo.)

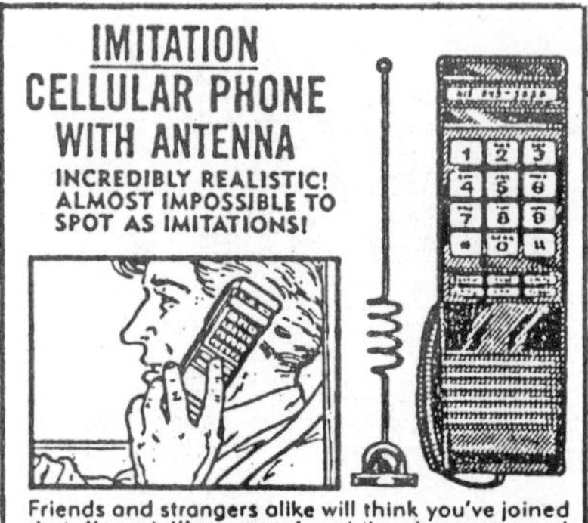

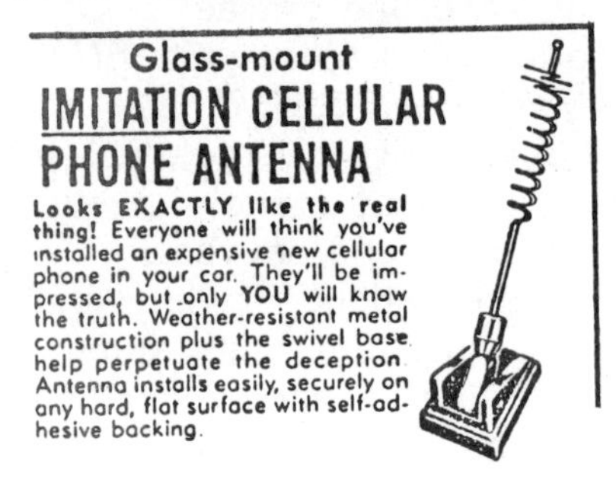

Fake CMT's and antennnas, as well as lookalike CB sets have ridden in on the wave of CMT's popularity. These items are from the Whitney automotive supply catalog.

CMT equipment. These are big sellers, too!

All of this fuss is about a communications service that is already available to the public in all major and most intermediate sized cities, and is rapidly becoming available in communities with smaller populations. Portable and mobile CMT units are installed in cars, boats, trailers, motor homes, RV's, vans, highway trucks, taxicabs, jacket pockets, attache cases, limos-- any lots of other places. The fancier units offer every convenience of home telephones-- direct dialing to any telephone in the world, redialing, speaker phone, conference calling, hanset volume control, dialed number LED or LCD display, memory bank for frequently called numbers, even a lock to prevent unauthorized use.

Calls to/from CMT's sound just about like regular landline calls, most folks receiving a call from a CMT don't realize

A CMT installed in an attache case naver fails to generate interest. This one comes from Spectrum Cellular Corp.

that they aren't talking to a regular landline telephone. The person with the CMT knows the difference, however, because they've probably paid anywhere from $500 to $2500 for their CMT installation, plus a monthly service charge to have a CMT account with a telephone company (so-called "wireline" CMT service), or an independent CMT service ("non-wireline") supplier. The CMT owner also pays, by the minute, for all incoming as well as outgoing calls. With insurance, etc., this could easily average well over $250 per month in some areas of the nation. Obviously, it's worth every penny to those who need to stay in touch while they're on the go-- and can afford the tariff of owning a CMT as a necessity or luxury toy.

The concept upon which CMT service operates requires a cluster of transceivers located in various localized zones ("cells"). Each cell has its own transmitting/receiving site, connected by landline telephone to the company's central Mobile Telephone Switching Office (MTSO). Computers at the MTSO monitor all on-the-air activity taking place in each of

THE DIRECT LINE™
Controls and Indicators
Dial In Handset Model

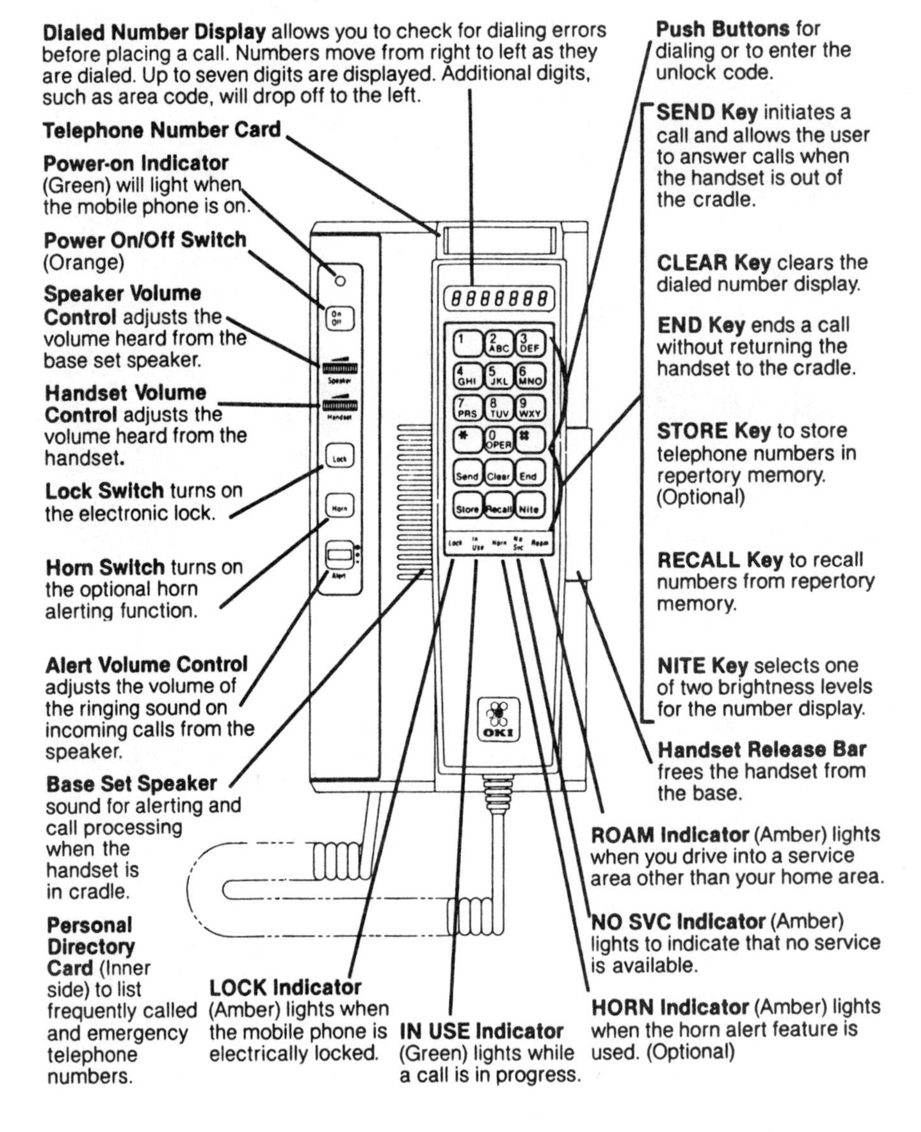

The features and controls on a deluxe CMT (Courtesy OKI).

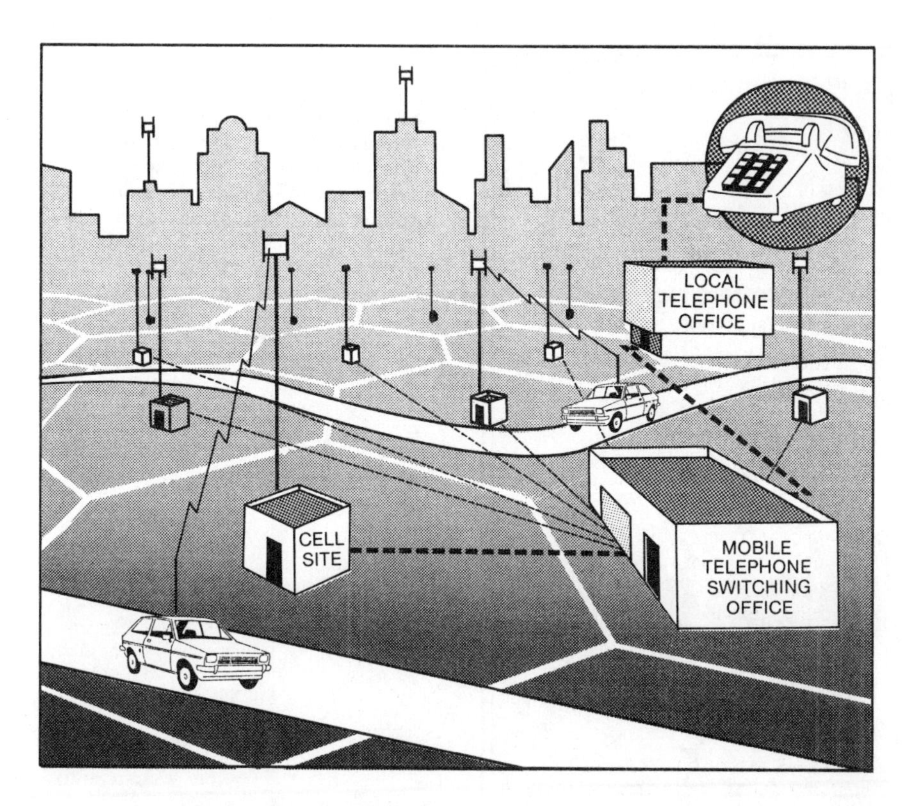

This diagram shows the basic layout of a typical CMT system, and how a vehicle passes through various cells as it travels.

the cells. As a moving vehicle transits out of the operating range of one call, and into the operating zone of the next adjacent cell, the MTSO switches ("hands off") the conversation to the facilities of that cell. Although this involves a change in both base/mobile operating frequencies, those engaged in conversation over the CMT are unaware that anything has taken place.

The MTSO is, of course, connected to the landline telephone service in the community. Each CMT is programmed with an individually assigned number that can be automatically read by the MTSO. Roamer (out-of-home area) CMT service is also available to most customers who make arrangements for that service.

CMT operations take place in the 800 MHz band, and base stations (individual cell transmitters) repeat the mobile

This is what most mobile cellular antennas look like. The corkscrew in the center is a typical design feature of CMT whips. This one is roof mounted, but they also come in trunk-lip, rear deck, mag mount, and thru-glass varieties.

transmissions so both sides of a conversation are heard when monitoring a base station. Each cell site is equipped for operation on a large number of channels so that numerous simultaneous conversations can take place in the same local area. CMT channels (all 830+ of them) are spaced every 30 kHz, and occupied in each local area by one wireline and one non-wireline company's signals. To persons tuning the band, it won't make much difference as to which-is-which, they both send out signals that are identical in every respect. The cell cites (base stations) used to operate between 870.0 and 890.0 MHz, but the band has now been expanded to 869.0 to 894.0 MHz in order to provide 5 MHz more worth of additional channels. Mobile units all operate between 824.0 and 849.0 MHz, but use low power. Both sides of the conversation can usually be heard by monitoring only the base station channel.

Mobile channels are used by cars, boats, trains, portables, and fixed station units.

Cellular handhelds (like this one from Radio Shack) and also car phones are now being used during surveillance and stakeouts by federal agencies as well as state, county, and local law enforcement agencies. The cellular mobile antennas are less conspicuous than standard VHF/UHF antennas, and cellular frequencies appear to provide more privacy than the agencies' regular VHF/UHF communications frequencies. This is because cellular communications are relatively short-range, there are fewer scanners in use that can pick up the 800 MHz band than other bands, and also because most people don't realize these agencies are using cellular phones for this type of undercover work. Indeed, in several major metropolitan areas, some of the most sensitive intelligence, investigative, and enforcement operations run virtually all of their communications via cellular telephones!

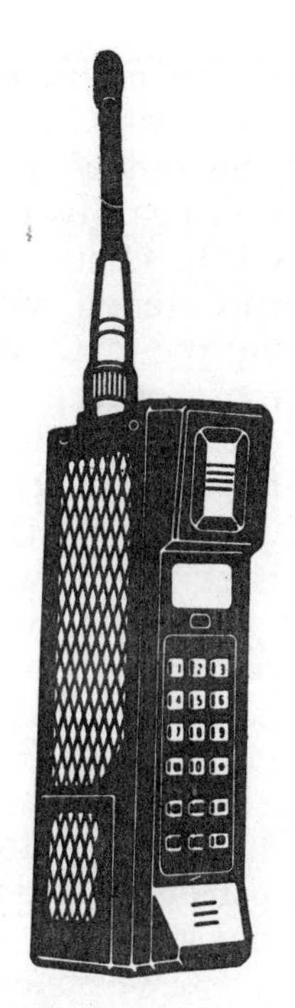

A scanner set at 30 kHz increments between 870.00 and 890.00 Mhz (or 870.00 and 896.00 MHz), will find stations in every area where cellular service exists. Conversations, if they take too long, will abruptly cut off in mid-word when the mobile unit leaves one cell site and enters another. Often, it's possible to continue search/scanning the band and locate the continuation of the same conversation as it takes place via the other cell site facility on another frequency-- assuming that you're in receiving range of that cell site.

A person with a scanner has no way of knowing which exact cells are being monitored during any given exchange of transmissions, it makes little difference. It's a random deal-- just hit-and-miss, but in most areas there are so many CMT conversations going on that a scanner set in search/scan mode will find no shortage of chatter to bring forth.

In addition to just plain folks using CMT, all sorts of other not-so-plain people are there too. Mixed in with doctors, salesmen, lawyers, and routine business and family yakking, there's more-- lots more! Daytime usually provides no end of general gabbing, but at night the CMT channels take on a completely different image.

At night, especially, the language (and often the topics) are strictly XXX, and not for persons of delicate sensitivities. Apparently neither the FCC nor the companies providing CMT services give a damn about the absolutely raunchy language on CMT channels. Broadcasting stations have been fined, or threatened with the loss of their licenses, for lesser utterances on the air. Moreover, the CMT service is being used for the most outrageous purposes. Drug deals are openly discussed, and one newspaper reporter wrote a column about having monitored arrangements for an apparent mob "contact." Major users of CMT's at night include well-heeled married romeos checking in with their girlfriends behind wifey's back.

It's no wonder the CMT industry pushed so hard for the passage of the ECPA, they certainly were less than anxious for continued public interception of such use of the airwaves. Down deep, of course, they wanted to be able to offer cellular service by assuring the prospective users of the service that there were federal laws in effect that assured communications privacy-- even though the privacy, from a practical standpoint, doesn't exist to any degree whatsoever!

CMT's are also used by law enforcement agancies for surveillance work, and that includes federal agencies as well as local departments.

Because of the wide assortment of diverse conversations that go out over CMT's, some of those who have been known to monitor the frequencies include law enforcement and intelligence agencies, drug dealers, those who rip off drug dealers, blackmailers, private investigators, people looking for hot stock market tips, wives/husbands who suspect that their spouse is cheating, representatives of foreign governments, persons doing industrial espionage, and a massive army of casual listeners, hobbyists, snoops, yentas, and busybodies.

If anything, the ECPA has backfired! It has given CMT subscribers the false impression that their calls are private when, in fact, they might as well be holding their conversations over public address systems installed on the busiest corner in town. Therefore, they talk about things that they might not have discussed had they not been deceived into having an illusion of privacy. Also, the ECPA has created a curiosity as to what the hell is going on over these frequencies that nobody's supposed to hear. Made people take a listen who would have otherwise not given a damn.

Driver Charged In 4-Car Crash

A driver talking on a car telephone ran a red light Tuesday in Clarence, hit another car broadside, pushing it into two other cars, state police said.

Frank J. [illegible], 40, of [illegible] Constitution Ave., West Seneca, was charged with running a red light and failing to wear a seat belt after the accident at 10 a.m. on Wehrle Drive at Harris Hill Road.

Troopers said Gregory L. [illegible], 22, of [illegible]ville Road, East Aurora, the driver of the car struck first, suffered a possible broken leg and was admitted to Millard Fillmore Suburban Hospital, Amherst.

Wailand was treated for a head injury, and the other two drivers, Robert [illegible], 31, of South Irwinwood Road, Lancaster, and Janet [illegible] of Vanderberg Drive, Lancaster, did not require treatment.

One of the hazards of CMT operation is yakking while in motion. More than one driver has crashed while deeply engrossed in a CMT conversation.

From the extremely personal and intimate conversations going out over CMT's, the violent arguments, sleazy business deals, and other tacky chatter it's quite apparent that folks don't really appreciate the fact that their conversations can be so easily heard by outsiders. Some CMT owners have told me that they are absolutely certain that their conversations are completely protected-- either the person who sold them their equipment told them that, or they have simply assumed that such was obviously the case. Other people may know or suspect that their CMT conversations are an open book, and actually be careful about what they say for the first couple of minutes-- but CMT conversations are so much like landline conversations that in short order the people forget to restrain themselves. Fact is that most of the time, the CMT user doesn't even bother to notify the person to whom they're talking that the call is going out on the airwaves via CMT, so the landline person doesn't even realize the situation.

Of course, the smart folks buy a voice scrambler for their CMT and a matching one for whatever landline phone they normally call with chatter they don't want overheard. A unit like the one from AMC Sales, Inc. (9335 Lubec St., Box 928,

Downey, CA 90241) has 13,000 selectable codes. Each unit (two are required) costs $369; a small price to pay for assured security based upon some of the tacky deals the high rollers feel the need to discuss over their CMT's. Still, the majority of CMT conversations are in the clear.

Monitoring CMT's, despite the unenforceable ECPA, has become a widespread practice for any rumber of amateur and professional reasons. Illegal? Yes-- most definitely! But there are so many pieces of equipment capable of receiving CMT frequencies already in the hands of the public, and the public has tasted the forbidden delights of hearing what goes on there that eavesdropping on CMT's has become an undeniable fact of life. From those to whom I've spoken about the practice, the rationale is that people selling the drugs over the CMT's are the law breakers, not those who listen in awe to their activities.

Of course, the CMT industry plays down or totally ignores the fact that their prize baby has such a darker side. They would prefer that CMT's were perceived as being mainly used for calling ahead to confirm a haircutting appointment, or ordering a pizza. Yet, according to a story by David Enscoe in the Ft. Lauderdale News and Sun-Sentinel (a Florida newspaper), the commander of the Broward County Sheriff's narcotics squad said, "A cellular phone is a great tool for drug traffickers but for us its a killer. It's the biggest hurdle we've run into."

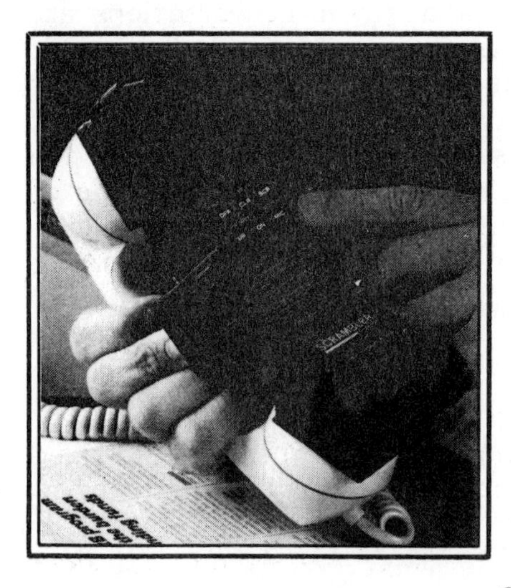

For those who want privacy, a voice scambler is the only way to go, yet the majority of CMT conversations go out "in the clear." (Courtesy AMC Sales, Inc.).

Drug rings

Cellular car phones are driving agents buggy

By DAVID ENSCOE
Fort Lauderdale News & Sun-Sentinel

WEST PALM BEACH, Fla. — Tapping the telephone at Ernesto Benevento's West Palm Beach home didn't help federal agents crack one of the largest heroin-smuggling rings in history last year.

Benevento was doing his business by cellular car phone.

"We had a tap on his home phone for three or four months, but he wasn't using it," said John Carroll, a U.S. attorney in New York. "He was using the cellular phone in his car. We were unable to technically do the intercept."

Actually, the technology does exist to intercept cellular conversations, but it is a complicated and expensive procedure.

"We have the technology to intercept beeper signals and cellular phones," Carroll said. "We can intercept every phone transmission in the country. But it's not just technology. It's the cost of doing it."

Car phones now are standard equipment for south Florida's big-time drug dealers, authorities say.

"Everybody we deal with has them. I mean everybody," said Lt. Ron Cacciatore, who commands the Broward County sheriff's narcotics squad. "A cellular phone is a great tool for the drug traffickers, but for us it's a killer. It's the biggest hurdle we've run into."

Capt. Tom Thompson, head of the Palm Beach County sheriff's organized-crime bureau, said, "We've found them almost every time we've made an arrest in the last year."

Drug agents say the high-tech cellular phones make it more difficult to keep tabs on the traveling drug salesman. The Benevento investigation is a case in point.

"Initial efforts to do interception were hampered by the fact that the guys were using cellular phones," said Bill Simpkins, a Drug Enforcement Administration agent.

Ironically, it was high technology that helped authorites break up the international heroin ring and put Benevento and his associates in jail. He kept all records of his criminal dealings in a personal computer.

"The phone tap wasn't productive, but the computer records . . . is what made our case," Carroll said.

Companies that sell cellular phones acknowledge the problem.

"It's a big concern for us, but there isn't a whole lot we can do to stop it," said Jim Earle, a spokesman for BellSouth Mobility, the largest of the two cellular-phone companies serving south Florida.

Earle said drug agents regularly subpoena phone records to aid in criminmal investigations, and "we work closely with them."

Catching drug dealers can help the phone company as well as the law. According to Earle, the same people who use cellular phones to deal drugs try to avoid paying their bills, often by altering the phone's electronic identification numbers.

Cellular phones are computer-controlled radios, but conversations on cellular phones are more difficult to bug than radios. Cellular conversations can be assigned to one of thousands of frequencies that may change as the car moves from one place to another.

Signals from cellular phones are beamed to specific "cells," which relay the signals. Each cell has only a limited range, so when a car drives out of the range of one cell, another cell takes over, and a different frequency is used.

"You never know what cell site the radio wave is going to," Cacciatore said. "It all depends on which cell he's closest to. It makes it very hard for us."

Drug dealers and other tacky types are major users of CMT's.

Capt. Tom Thompson, head of the Palm Beach County Sheriff's organized crime bureau, told Enscoe, "We've found them almost every time we've made an arrest in the last year."

Bill Simpkins, of the DEA in Florida, observed, "Initial efforts to do interception were hampered by the fact that the guys were using cellular phones."

Enscoe's story pointed out that the largest heroin smuggling ring in history made its deals via CMT. Regular hardwired phone taps were therefore useless. The drug agents had to subpoena CMT billing records and then try to figure out who their contacts were. CMT is therefore a totally unique and most useful tool for the illicit drug indistry which is one of the largest groups included in those whose privacy the ECPA was intended to assure.

Oddly enough, the sleazeballs have little enough interest in looking very kindly upon the CMT industry that has provided such a wonderful tool for all manner of shady doings. Seems that there is a thriving underground market in "chips" that go into CMT and provide bogus, misleading, unassigned, or otherwise invalid CMT unit billing numbers in order to rip off the Common Carriers providing the CMT service. The fake ID's are automatically sent out over the air by the car phones, and duly recorded by the computer at the MTSO-- but when the time comes for a bill to go out, lots of blinking red lights and bells go crazy. The FBI has traced many of these wiseguys by contacting those whose numbers they called, but it does seem that this problem will be an ongoing nuisance.

That's the CMT in a nutshell. If you thought that the CB radio of the mid-1970's was a looney bin, you ain't heard nothin...yet!

4 4 4 4 4 4 4 4 4 4 4 4 4 4 4 4 4

IMTS (Non-Cellular) Telephone Service

Mobile telephone service has been available to the public since the mid-1940's when channels were first established in the 35 and 152 MHz bands. Service in those early days was very basic, the mobile subscriber was assigned to use one specific channel, and it calls from mobile units were made by raising the operator by voice and saying aloud the number being called.

Mobile units were assigned distinctive telephone numbers based upon the coded channel designator upon which they were permitted to operate. A unit assigned to operate on Channel "ZL" (35.66 MHz base station) might be ZL-2-2849. The mobile number YJ-3-5771 was a unit assigned to work with a Channel YJ (152.63 MHz) base station. All conversations meant pushing the button to talk, releasing it to listen.

As the years passed, most of these systems blossomed and became more sophisticated, adding channels, installing equipment that provided subscribers with automatic dialing, ringer service, multi-channel operation, and similar. By the beginning of the 1970's, these services had, for the most part, become something so different than what they had been a couple of years earlier that they became known as Improved Mobile Telephone Service (IMTS), incorporating frequency pairs in the 35, 152, and 454 MHz bands (by 1988, the old 35 MHz--Z-channels were all reassigned to radio paging duties, along with their associated 43 MHz mobile frequencies).

Of course, not all current 152 and 454 MHz systems offer identical modernization, as there are a wide variety of sophistication levels offered by individual companies.

At the lowest level of modernization, there are still companies providing service for sets with push-to-talk buttons where the operator must be told what number is being called.

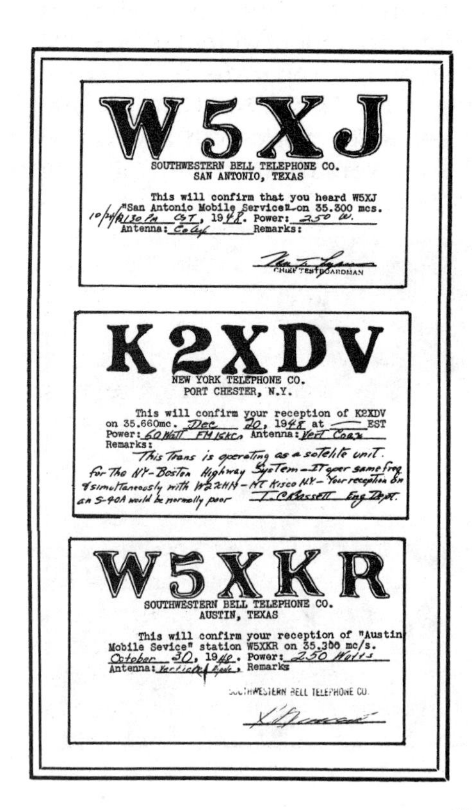

W5XJ
SOUTHWESTERN BELL TELEPHONE CO.
SAN ANTONIO, TEXAS

This will confirm that you heard W5XJ "San Antonio Mobile Service" on 35.300 mcs. 10/24/48 1:30 PM CST, 1948. Power: 250 W.
Antenna: Coax Remarks:

CHIEF TESTBOARDMAN

K2XDV
NEW YORK TELEPHONE CO.
PORT CHESTER, N.Y.

This will confirm your reception of K2XDV on 35.660mc. Dec 20, 1948 at ____ EST
Power: 60 Watt FM 15KC, Antenna: Vert Coax
Remarks:
This Trans is operating as a satelite unit for The NY-Boston Highway System--It oper same freq & simultaneously with W2ZHH--Mt Kisco NY--Your reception on an S-40A would be normally poor
T. C. Bassett Eng Dept.

W5XKR
SOUTHWESTERN BELL TELEPHONE CO.
AUSTIN, TEXAS

This will confirm your reception of "Austin Mobile Sevice" station W5XKR on 35.300 mc/s.
October 30, 1948. Power: 250 Watts
Antenna: Vertical Coax, Remarks

SOUTHWESTERN BELL TELEPHONE CO.

In 1948, when car phones were still considered experimental, the author logged these three telco bases on the original 35 MHz "Z" band channels. W5XJ and W5XKR were in San Antonio and Dallas, TX and ran 250 watts each. K2XDV was a local 60 watt base used for the New York-Boston highway system. All three stations verified reception with these prepared reply cards-- decades before the ECPA was devised.

One step above this rather primitive type of service, are those companies set up for mobile units with dial-type access to landline telephones-- rather than the pushbutton/tone type units used with more sophisticated modern systems.

Some companies aren't set up for dealing with "roamer" units-- these are mobile units operating outside of their own home areas seeking to make or receive car phone calls. In other words, IMTS mobile phone service, on a national basis, is somewhat of a mixed bag of diverse services in which the mobile units from one area are not always suitable for placing/receiving calls through the facilities of Common Carriers in other areas. And, to further confuse matters, while all U.S. channels are available in Canada, some 14 channels used in Canada still aren't available in the United States.

Nevertheless, the system, as a whole does function well and there are still new facilities being added on these bands. This, despite the fact that the advent of 800 MHz cellular service in the early 1980's has generally caused 152 and 454 MHz IMTS service to become somewhat of an odd stepbrother to the newer, more fashionable and "in" cellular service. Our listings, therefore, reflect the available services as we believed them to be at press time. New stations do come on the air from time to time, and we'd appreciate hearing from our readers about such. This is especially true on the 454 MHz frequencies, including those used for aircraft telephone calls.

IMTS base stations usually offer two-way service for about 20 miles out, limited only by the call-in distance of the mobile units (30 watts maximum on VHF, 25 watts maximum on UHF). Signals from the base stations can actually be copied at distances much further than 20 miles. Inasmuch as the base stations repeat the transmissions of the mobile units, so a scanner tuned to a base station's channel would pick up both sides of any conversation.

One of the peculiarities of many IMTS Common Carrier services is that there is a constant 2 kHz audio tone transmitted at all times when the channel isn't in active use with a call. The tone is sent out to guide multi-channel mobile transceivers to an available channel. It's annoying to listen to for any length of time, moreover scanners lock up on these tones and will refuse to continue scanning when they encounter one. A few old Bearcat scanners had a built-in switchable 2 kHz audio filter that would defeat this problem, but modern

The more sophisticated modern IMTS car phones look basically similar to cellular phones. In fact, the cellular phones were based upon IMTS phones. Better grade IMTS units permit direct dialing to every telephone in the world, just likle cellulars-- but the overall cost of having IMTS totals up cheaper than a cellular. In some areas, IMTS is still the only game in town when it comes to car phones.

scanners aren't equipped with such filters. It's probably possible for a communications technician to install a filter that will permit the scanner to ignore the tone and continue to scan when one is encountered.

Aeronautical service on the 454/459 MHz channel pairs seems to be expanding at a rate faster than the mobile services listed in this section. New ground stations are going on regularly. The aircraft stations (on 459 MHz) can be copied from considerable distances out-- perhaps several hundred miles, depending upon the altitude of the aircraft.

IMTS has really become almost a forgotten service. The channels in any given area soon become less active upon the local availability of cellular telephone facilities. Ask the average person about getting a car phone and they'll immediately assume you mean a cellular unit. Even the telephone company gives IMTS the low profile treatment-- if you ask telco about a car phone they try to point you towards cellular. You have to specifically ask for information on IMTS, which they also refer to, in a rather disdainful way, as "non-cellular" mobile telephone service.

But IMTS is still there, anyway-- serving the many

customers in areas that don't yet have cellular service, plus many subscribers in cellular areas that either had IMTS before cellular came along, or who simply prefer IMTS to cellular for reasons of their own. IMTS service is also used as a fixed service for providing telephone access to remote rural homes, farms, hunting lodges, logging camps, oil fields, mining camps, fishing camps, and other customers beyond the reach of landline facilities.

In the station listings, channels indicated with the letter "A" in front of a number (A8, A12, etc.) are aeronautical telephone call channels. Note that aero channels 1, 2, 3, 4, 5, and 6 are also known (respectively) as Channels QM, QX, QL, QW, QH, and QS, although we don't use those designations in this listing.

In the U.S., the ECPA prohibits monitoring IMTS stations, except the ones providing aeronautical telephone service.

Standard IMTS Channels *= Canada only.

Channel	Base Freq.	Mobile Freq.
JJ*	152.48	157.74
JK	152.78	158.04
JL	152.51	157.77
JP	152.57	157.83
JR	152.81	158.07
JS	152.69	157.95
JW*	152.84	158.10
XJ*	152.495	157.755
XK*	152.525	157.785
XL*	152.555	157.815
XP*	152.585	157.845
XR*	152.615	157.875
XS*	152.645	157.905
XT*	152.675	157.935
XU*	152.705	157.965
XV*	152.735	157.995
XW*	152.765	158.025
XX*	152.795	158.055
XY*	152.825	158.085
YJ	152.63	157.89
YK	152.66	157.92

YL	152.54	157.80
YP	152.60	157.86
YR	152.75	158.01
YS	152.72	157.98
QA	454.45	459.45
QB	454.55	459.55
QC	454.375	459.375
QD	454.425	459.425
QE	454.475	459.475
QF	454.65	459.65
QJ	454.40	459.40
QK	454.525	459.525
QO	454.575	459.575
QP	454.50	459.50
QR	454.60	459.60
QY	454.625	459.625

Aeronautical Telephone (A#) Channels

(All ground stations send signal tones on 454.675 MHz)

A1	454.95	459.95
A2	454.90	459.90
A3	454.85	459.85
A4	454.80	459.80
A5	454.75	459.75
A6	454.70	459.70
A7	454.725	459.725
A8	454.775	459.775
A9	454.825	459.825
A10	454.875	459.875
A11	454.925	459.925
A12	454.975	459.975

Non-Cellular Mobile Telephone Directory

Alabama

Anniston	JR
Arab	YL
Atmore	JL YK
Birmingham	JK YJ YK YP YR YS QC QK
Camden	JS
Decatur	JR
Dothan	YJ YK QC QD
Foley	YJ YL
Gordon	JK
Goshen	JL YS
Huntsville	JK YL
Lanett	JL YS
Leesburg	YR
Mobile	JK YP YR YS
Monroeville	JP
Montgomery	YJ YP YR
Oneonta	JP
Pell City	JL
Sheffield	YS
Troy	A10
Tuscaloosa	JR
Union Springs	JP

Alaska

Fairbanks	A5 A6
Kenai	YR

Arizona

Chinle	YL
Flagstaff	JL YR
Grand Canyon	A12
Page	JS
Parker	JL YJ
Phoenix	JK JL JP JR YK YK YL YP YR YR RS QA QB QC QD QE QF QJ QK QO QP QR QY A8
Sierra Vista	YS
Tucson	JK JR JS YJ

Williams	YP
Yuma	YK YP YR YS

Arkansas

Alma	JL
Bald Knob	YK
Blytheville	JK
Booneville	JP
Bull Shoals	YL
Clarendon	YS
Clarksville	JR
Conway	JP
Crossett	YP
Danville	YS
De Queen	JK
Dumas	JR
El Dorado	JL YR
Elaine	JS
Fayetteville	YS
Forest City	YL
Ft. Smith	YJ YK YR
Harrison	JP
Hope	YJ
Jonesboro	YR
Lewisville	JS YS
Little Rock	JL JR JS YR QE QK A6
Marked Tree	JK JS
McCrory	JK
McGehee	JP
Mountain Home	YK
Mountain View	YP
Newport	JL
Paris	YP
Pine Bluff	YK YL
Prairie Grove	JK
Redfield	YK
Russellville	JL YL
Star City	JK
Stuttgart	YP

California

Alturas	YR

Bakersfield	JK JR JS YJ QO
Barstow	YP
Big Bear Lake	JS
Blythe	JR YS
Boron	YL
Burney	JS
Chico	JK YJ
Chualar	JL YJ QE QJ
Clearlake Oaks	JL
Coalinga	YP
Colfax	JR QK
Colusa	YP YS
Corcoran	QP
Courtland	QR
Covina	QR
Dos Palos	JK
Elk Grove	QY
Eureka	JR YJ
Exeter	YK QJ QK QY
Fall River Mills	JL
Foresthill	YL
Ft. Bragg	JS
Fresno	JL JR JS QO A3
Garberville	YS
Gilroy	YL
Hemet	YS
Imperial	JL YJ
Indio	JP
Kerman	YR YS
Kernville	JP
Lk. Isabella	JP
Lancaster	YK
Lompoc	JK
Long Beach	YK YR QF
Los Angeles	JK JL JP JR JS YJ YS QA QB QJ QO QP QY A4 A7 A10
Los Gatos	JK QE QF QK QO QR
Lucerne Valley	YL
Manteca	YL QE
Marysville	JP YK YR
Merced	YJ YP
Modesto	JL JS QK QO

Morgan Hill	YP
Novato	JP
Oakland	JR YK YR QJ QP
Oxnard	YP
Palm Springs	JS
Palmdale	JL
Patterson	YR QA
Pioneer	QP
Pomona	QO
Redding	JP A6
Redlands	QY
Reedley	YJ
Ridgecrest	YJ
Riverside	YL YP
Roseville	QC
Sacramento	JK JL JS YJ QJ
San Bernardino	JR
San Diego	JL JP JR JS YJ QB A9
San Fernando	QC
San Francisco	JL YJ YL QB A1 A8
San Jose	JP JS YS QK QY
San Luis Obispo	YL
Sanger	JP
Santa Ana	YL YP QE QK
Santa Barbara	YL QR A5
Santa Maria	YK
Santa Rosa	JK JS YS
Shingletown	YL
Stockton	JP YS QB
Susanville	YS
Taft	YR
Tahoe City	JL
Tulare	YL QB
Tulelake	YS
29 Palms	JK
Ukiah	YJ
Vallejo	YP QA
Ventura	JL YJ
Victorville	JK JP YJ YS
Vista	JL YL YP
Weaverville	YK
W. Los Angeles	QD QK
Willow Creek	YP

Yreka	JS
Yucca Valley	YJ

Colorado

Alamosa	JS YJ
Boulder	JP
Byers	JR YK YP
Colorado Springs	JP YJ YL YS QD QF QP QO
Cortez	JS YJ
Denver	JK JL JR JS YJ YL YP YR QA QB QC QD QE QF QJ QK QO QP QR QY A7
Eagle	JS
Eckley	YK
Ft. Collins	QA QR
Ft. Morgan	JL JP
Grand Junction	YR YS A4
Greeley	YK YS
Holyoke	JP YP
Hotchkiss	YP
Joes	JL
La Junta	JL
Pleasant View	YS
Pueblo	YP YS
Rangeley	YR
Sterling	JS YR
Trinidad	A10

Connecticut

Bridgeport	JR
Hartford	JL
New Haven	YJ
Stamford	YP
Waterbury	JK

Delaware

Dover	JS YR
Georgetown	JK YP
Wilmington	JK JP YP

District of Columbia

Washington	JK JL JR JS YJ YP YR YS QB QC QD QE QF QK QO QY A1

Florida

Avon Park	YS
Belle Glade	JP YR
Clearwater	JP YL YP YS QF QO QR QY
Cocoa	YK A3
Daytona Beach	JP YL
Ft. Lauderdale	JS YL QE QF QK QO QY
Ft. Myers	JP JR JS YJ YP
Ft. Pierce	JR YJ YL
Ft. Walton Beach	JL YJ YK
Gainesville	YL JS
Havana	JL
Homestead	JR JS
Jacksonville	JL JP YJ YK YP YR
Lake City	JK
Leesburg	JS YK YP
Live Oak	YS
Marianna	JR
Melbourne	JS
Miami	JK JL JP YJ YK YP YR YS QA QB QC QJ QP QR A7 A8
Naples	YK YS
New Port Richey	JK YR
Ocala	JK YJ YS
Okeechobee	JS
Orlando	JK JL YJ YR YS
Palatka	JR
Panama City	JS
Pensacola	JP JS
Perry	JP
Pt. Charlotte	JL
Pt. St. Joe	YR YS
Quincy	JL
Sarasota	JK JR JS YR QF QO QR QY
Tallahassee	JS YJ YK YL YP
Tampa	JL JS YJ QA QB QC QD QE QJ QK QP A5

Wauchula	YP
W. Palm Beach	JK JL JP YK YS
Windermere	YP
Winter Haven	JP JR YJ YL

Georgia

Albany	YL
Atlanta	JL JR JS YJ YK YL YP YR A7 A8 A9
Augusta	JL
Blakely	JS
Blue Ridge	YL
Brooklet	JL
Columbus	YJ
Dalton	JK QP
Ellijay	YS
Fairmount	JP
Folkston	YL
Glennville	JP
Hawkinsville	JL
High Point	YL
Hinesville	JK YR
Lafayette	JS
Macon	YJ
Newington	YL YP
Omega	YJ
Plains	YP
Reynolds	JS
Ringgold	YK
Savannah	YJ
Statesboro	JR JS
Thomaston	YK
Twin City	YK YS
Vienna	YR
Washington	JP
Waycross	A1
W. Brow	QJ
W. Point	JL YS

Hawaii

Hilo	YK YP
Lihue	YK

Oahu	JK JL JR JS YJ YR
Wailuku	JP

Idaho

Albion	YK
Boise	JL JP JR YR A4
Coeur d'Alene	JS YR
Filer	YL YR YS
Idaho Falls	JS YJ A10
Lewiston	JS
Mc Call	JK YS
Moscow	JR
Pocatello	YL YR
Rupert	JK JL JP YP QB QC QD QP QR
Twin Falls	JR JS YJ

Illinois

Alton	A4
Aurora	YK
Bloomington	JL JR YS
Brownstown	YS
Canton	JP
Carbondale	JL
Carthage	YJ
Casey	YP YR
Centralia	YR
Champaign	YP YR
Chicago	JK JL JP JR JS YJ YR YS QA QB QC QE QF QJ QK QO QP QR QY A1
Colchester	JS YP
Danville	JL
De Kalb	JK JR
Decatur	JP YJ
Dixon	YR
Effingham	YJ
Elgin	YP
Freeport	JP YK
Galesburg	JL YS
Golden	YS
Gridley	QC
Harrisburg	JR

Jacksonville	JR YJ
Joliet	YL YP
Kankakee	JR
Lincoln	YK
Litchfield	YL
Louisville	JP
Mattoon	JS YL
Mendon	JK
Mt. Vernon	YJ
Olney	JK YK
Ottawa	JL JP
Owaneco	JL
Pekin	JK
Peoria	JS YJ YL YP
Pontiac	YK
Princeton	YK
Quincy	YR
Rock Island	JS YJ TL YR
Rockford	JS YJ YL YS
Rossville	JP
Savanna	JK
Springfield	YP YR
Sterling	YP
Streator	JK
Utica	JL JP
Waterloo	YS
Watseca	YJ

Indiana

Anderson	JP
Atlanta	QF
Batesville	YP
Bloomingdale	JS YS
Bloomington	YK
Camden	JR YS
Clayton	YP
Cloverdale	JP
Columbus	JP
De Motte	YP
Elkhart	YP YR
Evansville	JS YJ

Fairmount	QJ
Ft. Wayne	JP JR YJ YL
Gary	YK
Greensburg	JK
Indianapolis	JK JR JS YJ YL YR QB QC QD QE QJ QK QO QP
Jasper	YP
Kokomo	JK
Lafayette	YL YR
Linden	QA
Madison	YK
Marion	YR
Markleville	YP
Maxwell	QA QO
McCordsville	YS
Monon	JS
Monrovia	QY
Muncie	YK
New Harmony	YR
Portage	YL
Richmond	JR
Rochester	JL
Rockport	JP
Seymour	JR YL
South Bend	JK JS YJ YS
Star City	JP YK
Swayzee	QY
Terre Haute	JR YL
Thorntown	YK
Vincennes	YS A11

Iowa

Bloomfield	JK
Brooklyn	YJ
Cascade	YK
Cedar Rapids	JL YR
Chariton	YP
Clear Lake	JS YK
Coon Rapids	JS
Cumberland	YK
Denison	JK
Des Moines	JK YJ YR YS

Dubuque	JS
Elk Horn	YP
Emerson	JR
Fairfield	YS
Ft. Dodge	JK JP YP
Gillett Grove	YR
Gowrie	YL
Harlan	YJ
Harper	JR
Havelock	YJ
Lake Mills	JP
Lawton	YJ
Lidderdale	YR YS
Manchester	YJ
Mt. Ayr	YS
Mt. Pleasant	YL
Newton	YL
Otter Creek	JP
Panora	JP
Postville	JP
Ringsted	YS
Rockford	YJ
Sanborn	JR YL
Schaller	JL
Sioux Center	JK JP YS
Sioux City	YL YP
Waterloo	JK YP A12
Wellman	YK
W. Bend	JS
W. Branch	JK
W. Liberty	JP
Wilton	YP
Woodward	JL

Kansas

Ashland	JR YP
Chanute	YP
Colby	A9 A11
Conway Springs	JR YS
Delevan	JS
Dodge City	YJ
El Dorado	YK

Ellinwood	JP
Emporia	YJ
Garden City	YP
Girard	JR YS
Grainfield	YP
Great Bend	JL YJ YR
Harper	YP
Haviland	JK YK YL
Hays	JK YK YL
Hutchinson	YK
Independence	YJ
Junction City	JK
Kendall	JP
Lawrence	JR
Lenora	JR YS
Leoti	JL YR
Liberal	YK YL
Long Island	JP
Manhattan	YK
Meriden	JP
Natoma	JS YJ
Newton	YP
Olpe	JP
Potwin	YS
Rexford	YK
Russell	JP YS
Salina	JL JS A1
Scott City	JS
Sharon Springs	YL
Topeka	YP
Tribune	JK YK
Udall	JP JS YL
Ulysses	JL YJ
Wellington	YK
Wichita	JK JL YJ YR

Kentucky

Ashland	JK
Bowling Green	YK YL
Cave City	YP YS
Florence	YS
Frankfort	JP JS

Irvington	JR
Lewisport	YK
Lexington	JK JL YR YK YL YP YR
Louisville	JK JL YJ YR
Middleboro	A5
Owensboro	JK YS
Paducah	JK
Pikeville	JP YR
Prestonburg	YJ
Winchester	YJ

Louisiana

Alexandria	JP JR
Bastrop	QJ QK
Baton Rouge	YJ YP QA QB QC QF QJ QP QR
B onita	YK
Buras	YP
Cameron	JR JS
Carlyss	YP
Collinston	JR
Delcambre	JR YP
Erath	JP JR QD QK
Franklin	YS
Gonzales	JK
Houma	JS YR
Jennings	YJ YR
Lafayette	JL JS YK YL YR QA QC QF QJ QO QP QR QY
Lake Charles	JL JP YR QA QJ QO
Larose	YL QD QE QO
Leeville	JK QC QY
Monroe	JR JS YJ YL
Morgan City	JL
New Orleans	JL YJ YR YS A3
Plain Dealing	JK JR
Port Sulphur	JP
Shreveport	JL YJ YL QC QD QE QC QO QY A5

Maine

Augusta	JR

Bangor	YJ A1 A7
Houlton	YJ
Lewiston	JP
Norridgewock	YK
N. Anson	YP
Portland	YJ
Presque Isle	JP
Rockland	JP
Stratton	YJ
Strong	JS
Unity	YL YS

Maryland

Annapolis	JP
Baltimore	JL JR YJ YR QA QJ QP
Chestertown	YL
Cumberland	YL
Easton	YJ
Frederick	YL
Hagerstown	YJ
Havre de Grace	YS
La Plata	YL
Oakland	YJ
Rising Sun	YK
Salisbury	YL

Massachusetts

Boston	JK JL YK YL YP QB QD QE QK QP QR A3
Brockton	YS QF
Hyannis	JS
Lawrence	JP QY
New Bedford	JP
Pittsfield	YS
Springfield	JP YK YP
Worcester	JR JS QO

Michigan

Adrian	YJ
Alma	JR
Alpena	YJ
Ann Arbor	JL YK QJ QP QR

Battle Creek	YK
Benton Harbor	JR
Cadillac	JK
Camden	JL
Cheyboygan	JS
Chesaning	JP
Chester	JP
Detroit	JP JS YJ YP YS QY QD QE QK QO QY A2
Donken	JP
Eckerman	YK
Flint	JL JS YP QC QK
Grand Rapids	JL YR
Hiawatha Forest	YL
Homer	JS
Jackson	JR
Kalamazoo	JP
Lambertville	YP
Lansing	JK YJ YP
Ludington	JL
Manistee	JP
Michigamme Forest	YP
Millington	YS
Monroe	JL JR YR QA QB
Mt. Clemens	JL YK QJ QR
Munising	JK
Muskegon	JR
Osseo	YR
Petoskey	JL
Pontiac	JR YL YR QA QB QF
Port Huron	JS
Saginaw	YJ
Sault Ste. Marie	YJ
Traverse City	JR YJ YK
Wallace	YR

Minnesota

Ada	JL YK
Alexandria	YP
Annandale	QK
Ash River Trail	JK YK
Audubon	JP

Bertha	JL
Blue Earth	JK YP YR
Bygland	QK QO
Choklo	JP JR
Clara City	JR YL
Clear Lake	JL YL
Comstock	JS YR
Dalton	JR YL
Deer River	YJ
Duluth	A2
Ely	JL YR
Fairmont	YK
Fertile	YS
Fisher	YJ YL
Fossum	QJ
Hackensack	YR
Halstad	JR
Holloway	JL YJ
Karkstad	QY
Lengby	JK YP
Malung	QK
Mankato	JL YL
Minneapolis-St. Paul	JK JP JR JS YJ YP YR A3 A11
Monticello	QA QF
Nevis	YL
New London	YK
New Prague	QO
New Ulm	YJ
Nokay Lake	JS YK YS
Pequot Lake	JK JP JR
Perham	YJ YS
Pine Island	YK YS
Red Lake Falls	JS
Remer	YP
Roosevelt	QO
Silver Lake	YS
Sleepy Eye	JP
Spicer	YK
Springfield	QF
Svea	YR
Turtle River	JL JR
Twin Valley	QJ

Wauconia	YK
Wannaska	QK
Wabun	QY
Westbury	JP
Worthington	JS

Mississippi

Bay Springs	YS
Bruce	YP YR
Decatur	YR
Gulfport	JP JS
Jackson	YL A2
Natchez	JL YR
Olive Branch	YK
Pascagoula	JL
Rienzi	JK
Sunflower	JL

Missouri

Bolivar	YJ
Boonville	YK
Branson	JL
Brookfield	JP
Bynumville	JS
Cape Girardeau	YK
Carrollton	YJ
Chillicothe	YK
Clinton	JR
Columbia	JL YR
Crystal City	YL
Doniphan	JP
Eldon	JP
Excelsior Springs	YP
Farmington	YJ
Farrelview	YS
Festus	YL
Fulton	YJ
Gray Summit	YP YS
Hannibal	YK
Harrisonville	JP
Jefferson City	JR YS

Joplin	YK
Kansas City	JK JL JS YJ YL YR QA QB QC QD QE QF A2 A8
Kingdom City	YJ
Kirksville	YR
Lebanon	YP YS
Lee's Summit	YK
Louisiana	YP
Malden	YJ
Marshfield	YL
Maryville	YJ
Moberly	YP
Nevada	YP
Pattonburg	YL
Perryville	YP
Piedmont	JS
Pilot Grove	JK
Poplar Bluff	YK
Rock Port	JP YP
Rolla	YJ
Sedalia	YP
Sikeston	YP
Springfield	JK YK
St. Joseph	YK
St. Louis	JK JL JP JR JS YJ YK YR QA QB QC QD
Sullivan	JP
Thayer	JR
Warsaw	YL
Wentzville	YL
West Plains	YS

Montana

Baker	JK YR
Big Timber	JP
Billings	JR JS YJ YR A9
Bozeman	YK
Butte	YP
Cabin Creek	YK YS
Chinook	JK
Circle	JL
Conrad	YK

Culbertson	YS
Cut Bank	JL JS
Fairfield	JR YJ
Fallon	JP
Glasgow	JS YS
Glendive	JS YJ A3
Glentena	YJ
Great Falls	YP YS
Havre	JL YJ
Helena	JS
Joplin	YL
Jordan	JP
Kalispell	JS YL
Miles City	YP
Missoula	JR YJ A7
Reserve	YL YS
Richey	JR
Scobey	JK YP
Sidney	JK YP
Sunburst	JP
W. Sidney	JP
Westby	YK
Winnett	JL
Wolf Point	YL

Nebraska

Arlington	A6
Alliance	A5 A12
Arnold	YS
Auburn	YK
Aurora	YR
Bassett	JL JR
Beatrice	JR
Benkelman	JK
Blair	JL YK
Blue Hill	QK
Burwell	JS YP
Clarks	YP
Columbus	JP
David City	YS
Doniphan	JR
Fairbury	YL

Falls City	JK YR
Geneva	JS
Grand Island	JK YL YK
Hartington	JR YS
Hastings	YJ YS
Henderson	QA QE
Kearney	JS YL
Kimball	JP YJ
Lincoln	JK JP YJ YP QF QO QR QY
Mc Cook	YJ
Nebraska City	YS
Norfolk	JS YR
N. Loup	YS
N. Platte	JL YP
Omaha	JS YL YR QA QB QE QK
Plattsmouth	QD
Scottsbluff	YR
Sidney	JL
Tecumseh	JL
Thedford	JK
Wahoo	QP
Wauneta	JR
York	JL

Nevada

Alamo	JK
Boulder City	YJ YR
Caliente	YK
Elko	A5
Fallon	JK YK YS
Garnerville	YL
Las Vegas	JK JL JP JS YK YL YP YS QA QB QC QD QE QF QK QO QP QR A6
Lovelock	JL
Panaca	YK
Pioche	YK
Reno	JR JS YJ YR A2
Sand Springs	JR
Stateline	YP YS
Tonopah	JS YJ
Winnemucca	YJ

Yerington	JP

New Hampshire

Contoocock	YS
Dover	JS
Hillsboro	YR
Manchester	YJ
New London	JP

New Jersey

Asbury Park	YK QJ QP
Atlantic City	QA
Belle Mead	YP
Belvidere	JL
Flemington	JK
Morristown	QE QJ QR
Newark	JR YL YR QD QE QO
Pleasantville	JP YS
Sussex	YL
Toms River	QY
Trenton	YL YS QC
Vineland	YK QK
Wildwood	QY

New Mexico

Albuquerque	JL JP JR JS YJ YR A5
Carlsbad	JS YK
Cottonwood	YL YP YR YS
Crown Point	YP
Farmington	JL JP
Hobbs	JL JR YP
Maljamar	JP JR
Silver City	A3
Tatum	JS YJ YK
Truth or Consequences	YP

New York

Albany	JL JP JS YJ
Buffalo	JL JP JR JS YJ YR
Clayville	YL
Coram	QD QE QO

Dexter	YS
Elmira	A5
Fulton	YK
Jamestown	YL YP
Johnstown	YK
Middletown	QJ QR
Mineola	QJ QP QR
New Woodstock	YS
New York City	JK JL JP JS YJ YS QA QB QC QF QK QY A6 A8
Newark	JP YJ
Newburgh	JP YJ YR
Norwich	JK
Poughkeepsie	JL JR JS
Pultney	YK
Riverhead	QC QF QK
Rochester	JK JR YL YP YR YS
Roscoe	JS
Selden	QD QE QO
Sidney	YK
Syracuse	JP JR JS YJ
Utica	JL YJ YR
Vernon	YP
Walton	YP
White Plains	YK
Whitney Point	YR

North Carolina

Advance	YP
Albemarle	YL
Asheboro	JS
Asheville	YJ
Biscoe	YP
Brooks	YP
Chapel Hill	JP YS
Charlotte	JK JL YJ YS QA QB QJ QP A2
Concord	YK
Durham	JK YL
Fayetteville	JL JR JS YL
Goldsboro	JK YS
Greensboro	JL
Greenville	YK YR

Harmony	YP
Hickory	JS YL YR
Jacksonville	JP YP
Kinston	JS YJ
Marshville	YR
Mt. Airy	YR YS
New Bern	JL
N. Wilkesboro	JP
Raleigh	YJ YP YR
Rocky Mount	JL JP JR YL A11
Roxboro	JS
Salisbury	JR
Sanford	YK
Southern Pines	YJ
Wilmington	YJ
Winston-Salem	YJ

North Dakota

Bismarck	YJ YP YR A1
Bottineau	JP
Carrington	JR YL
Cavalier	YK YR
Colfax	YP YR QD
Columbus	YL
Ellendale	YJ
Epping	JS YJ
Fargo	JS YR A4 A7
Hazen	YL YR
Keene	JL YK
Langdon	YJ YL
Manning	YR YS
Minot	JS YJ YR YS QD QE
Mohall	JK JL YK
Park River	JL JP JS YP
Parshall	JP
Roseglen	JR
Stanley	YP
Walhalla	YS
Ypsilanti	YR

Ohio

Akron	JP YS

Ashtabula	JK
Bellefontaine	YK
Bryan	YS
Canton	JL YS
Celina	JS
Chillicothe	YK
Cincinnati	JL YJ YR
Cleveland	JL JS YJ YR
Columbus	JL JR YJ
Dayton	JP JR JS A6
E. Claridon	JK YP
Elyria	YK
Greensville	YS
Hudson	JK YP
Kenton	YS
Lima	JL
Lorain	JR
Mansfield	JK YP
Marion	YR
Medina	YL
Middletown	YL
Sidney	YR
Springfield	YP
Toledo	JP JS YK
Van Wert	YP
Youngstown	JS YJ

Oklahoma

Ada	JP YR
Altus	YS
Alva	YS
Anadarko	YL
Apache	QP
Ardmore	JL YK
Atoka	YK
Bartlesville	YK
Blanchard	QK
Broken Bow	YL
Burns Flat	JS
Canadian	JR
Canton	QD
Capron	JP

Carmen	QE
Crescent	QY
Davenport	JP QB
Drummond	QR
Duncan	JK YK
Elk City	JL JP
Enid	JK JK JR YJ YK YL YR
Eufaula	YP
Gaymon	JP JR
Hennessey	JP QB
Hinton	QY
Hugo	YP
Keystone	YL
Kingfisher	JS QA QK
Lawton	YP
Lindsay	JS
Lone Grove	YK QC
Manchester	JS
Mc Alester	YJ
Mooreland	QA
Muskogee	JP
Newcastle	QB
Oklahoma City	JK JL JR YJ YK YL YR YS QC QD QE QF QJ QP QR A3 A12
Paoli	QA
Ponca City	YS
Pond Creek	QP
Poteau	YS
Roosevelt	JR
Seiling	JS QY
Shawnee	JS
Stillwater	YP
Sulphur	YL YP QO
Talihina	JL
Thomas	QP
Tulsa	JK JL JR JS YJ YR YS QC QJ
Valiant	JR
Vinita	YP
Warner	YL
Watonga	YJ QO
Weatherford	JK YJ
Woodward	YJ YL

Oregon

Albany	JS QF
Arlington	JP
Astoria	YJ
Baker	YL YP
Beaverton	QD
Bend	YJ
Blue River	JP
Burns	JK
Colton	YR QE QJ
Coos Bay	YL YR
Detroit	YL
Estacada	JR QC QP
Eugene	JL YJ YK QB QR
Florence	JL YJ YK
Glide	JS
Grants Pass	JP
Hood River	YP YS
Klamath Falls	YP A12
La Grande	JP
Lebanon	YS
Lincoln City	JR
Medford	JL YJ
Mt. Vernon	JR
Newport	JP
Pendleton	YJ A8
Philomah	JK
Portland	JL JS YJ YK YL QB QF QR QY
Redmond	JR
Roseburg	YJ
Salem	JP YP A3
Stayton	QK
Sunnyside	QA
The Dalles	YJ

Pennsylvania

Allentown	JP JR YK YS
Altoona	YJ
Bedford	JL
Birdsboro	QC QK
Butler	YS

Carlisle	YK
Chambersburg	JP YS
Connellsville	YS
Donora	YL
Ephrata	YK QD QE
Erie	YJ YS
Export	JP
Forest City	JR
Galilee	YS
Gibsonia	YK
Greensburg	JR
Hanover	YP
Harrisburg	JL JR JS YJ YL YR
Hazelton	YL YR
Indiana	YP
Johnstown	JS
Kittanning	JK
Lancaster	JP
Meadville	JP
New Bethlehem	YL
Oil City	YR
Palmerton	JS QD
Philadelphia	JL JR JS YJ YL YR QB QD QE QF QJ QO QP QR
Pittsburgh	JL JS YJ YR QC QD QE QJ A4
Reading	YP
Rochester	JR
Roseville	JS
Scranton	JK YJ YP
State College	YK
Washington	YP
Wilkes-Barre	JK YJ YP
Williamsport	JP
Yellow House	QA QY
York	JK

Puerto Rico

Aguas Buenas	JL JS YL YS
Cerno de Punta	JL JS YL YS
El Yunque	JL JS YL YS
Maricao	JP JR YJ YR
Monte del Este	JL JS YL YS

Rhode Island

Providence	YJ YR

South Carolina

Charleston	JS YJ A4
Chesnee	YP
Chester	YP
Columbia	JL YP YR QA QJ
Florence	JL
Greenville	JL JP YK YR
Greenwood	YJ YL YS
Inman	JK
Iva	JK YP
Kingstree	JP
Lancaster	JS
Laurens	QF
Lexington	YK
Manning	JR
Moncks Corner	YP
North	YS
Pelion	YR
Ridge Spring	JS JP
Rock Hill	JP
Scranton	YK
Spartanburg	JR JS
Sumter	YJ YL
Walterboro	YS
Williston	JR YJ

South Dakota

Beresford	YK
Brookings	JP JS
Corsica	JR
Dell Rapids	JR
Highmore	YS
Hitchcock	JK
Letcher	JL
Onida	YK YP
Pierre	A10
Sioux Falls	YJ YR

Tennessee

Bristol	YJ
Chattanooga	JL JR YJ QA QC QE QK
Clarksville	YS
Cleveland	YR
Collegdale	YP
Columbia	YP A12
Cookeville	QD QO
Dyersburg	JL
Greeneville	YL
Jackson	JP YJ YK YL
Johnson City	YS
Kingsport	JR
Knoxville	JL JR JS YJ YK YS
Lafayette	YR
Memphis	JL YJ YP YS QA QE QK QP QR QY
Millington	JP
Morristown	JK
Murfreesboro	YL
Nashville	JK JL JP JR JS YJ QA QC QE QK QR QY
Oneida	YL YR
Pikeville	JP
Smithville	YK
Tullahoma	YR
Woodbury	YS

Texas

Abilene	JL JR YK YK
Alice	JS YK YL
Amarillo	JK JR YS A6
Athens	YR
Austin	JL JS YJ YL YP YS
Bay City	JL YJ
Baytown	JS
Beaumont	YJ YL YS
Beeville	JP
Big Spring	JK YJ
Brownfield	YK YL
Bryan	JL YK

Bullard	JP
Canadian	JK YK YS
Canyon	JP
Carlsbad	JS YK
Cisco	JP
Cleveland	YK
Clifton	JS
Colorado City	YR
Columbus	YP
Commerce	YL
Conroe	JK YP QK QY
Corsicana	YK
Corpus Christi	JL JK JR YJ YP YR YS
Crockett	JK
Cuero	YJ
Dalhart	JK
Dallas	JL JR JS YJ YR QB QE QF QJ QR QY A4
Decatur	JS
Del Rio	JK YK
De Leon	YR
Denison	YS
Denton	YP YS
Dimmitt	YL
Dumas	YR
El Campo	JS
El Paso	JK JL JR YK YR
Encino	YP
Fairfield	JR YL
Flatonia	YK
Floresville	JR
Ft. Stockton	YK
Ft. Worth	JL YK YL QA QC QO QP
Freeport	YP
Freer	YP
Gainesville	JK
Galveston	YK YL
Ganado	JK YS
Graham	YP
Greenville	YK
Harlingen	YJ YR A3
Hebronville	JL

Hemphill	JR
Hempstead	JP
Henderson	JR
Hereford	JS YJ
Houston	JL YJ YR YS QC QK A1 A9
Hub	YK
Hull	JR
Huntsville	YJ
Irving	QD QK
Jewett	YP
Karnes City	YK
Katy	JR
Kerrville	JK YK
Kilgore	JK JL YL YS
Killeen	JK
Kingsland	JR YL
Kirbyville	YK
La Sara	JL
Lake Dallas	JP
Lamesa	JS
Laredo	JK JS YK YS
Lazbuddle	YP YS
Levelland	JP YS
Liberty	YP
Littlefield	JL
Livingston	JS
Longview	JS YK
Loop	QB QK
Lubbock	JR YJ
Lufkin	JL YL YP YR
Madisonville	YK
Maple	JK
Mc Allen	JP JR YK
Mc Camey	JS YP
Midland	YK YL YR YS
Milo Center	QJ QK
Mission	YK
Monahans	YR
Mt. Pleasant	YJ
Muenster	JR
Navasota	YL
Nocona	YS

Nubia	YP YS
Odell	YL JS
Odessa	JL JR YJ YP YS
Overton	YR
Ozona	YR
Palestine	JS
Patricia	QA QF QJ QY
Pearsall	JK JP
Pecos	JK YL
Perryton	JL YR
Plains	YS
Plainview	YK
Pleasanton	YS
Pt. Lavaca	YP
Post	YR
Punkin Center	JL YP
Quanah	JK
Ralls	YS
Ranger	YJ
Refugio	JS
Rosebud	JP
Rosenburg	YK
San Angelo	JK JP YL YP YS
San Antonio	JL JS YJ YL YP YR QJ QP A8
San Marcos	JK YS QC QO
Santa Anna	JS
Seguin	JP
Seminole	YR
Silsbee	JP
Skellytown	YJ
Snyder	YJ
Sonora	JR YJ
Spearman	YP
Stamford	YL
Stillman	JK JS YL
Stratford	JK YL
Sulphur Springs	JK
Sweetwater	JS A2
Tahoka	QC QD QO QR
Temple	YK
Texarkana	JP YK YP YR
Tulia	YR

Tyler	YJ YP
Union	YJ YP
Uvalde	YJ
Vega	JL
Victoria	JP YK YL
Waco	YJ
Waxahachie	YP
Westway	JS YJ
Wichita Falls	JL YJ YR
Winnie	JK
Woodville	YJ

Utah

Kamas	YL YP
Moab	JK YK
Monticello	JL
Moroni	JP YJ
Neola	JK JR YK YL YP YS QB QP QR
Ogden	YK YL A3 A11
Price	YR
Provo	YP YS
Randolph	JK JR
Richfield	A3 A11
Salt Lake City	JL JP JR JS YJ YR
Tremonton	YP YS
Vernal	JL JP JS YJ
Wendover	JR

Vermont

Burlington	JP
Ludlow	JK
Waitsfield	QO

Virginia

Charlottesville	JS YS
Edinburg	JL
Gum Tree	YR
Lynchburg	JK JP
Martinsville	YK YL YP
Newport News	JK JP JR
Norfolk	JK JL JR JS YJ YL YP
Richmond	JL JR YJ YL YP

Roanoke	YJ
Waynesboro	YL

Washington

Aberdeen	JK
Bremerton	QF QR
Cle Elum	YL
Cowiche	JS
Eatonville	QE
Ellensburg	JP YP
Ephrata	YS
Everett	YS QD QK
Forks	JP JR
Halls Lake	QJ QP
Kalama	JK YS
Kirkland	QO QY
Long Beach	YK
Longview	JP YP
Lynden	YL
Morton	YR
Moses Lake	YJ
Mt. Vernon	YP
Naselle	JL
N. Bend	YK
Olympia	JS
Omak	YR
Othello	JL
Packwood	YJ
Pasco	YL
Poulsbo	YR
Richland	YR
Seattle	JL YJ YL QA QC QP A1
Spokane	JL YJ A6
St. John	JP YP
Sunnyside	JK
Tacoma	JP YP
Uniontown	JK
Wenatchee	JK
Yakima	JR YK

West Virginia

Beckley	A3
Charleston	JR YJ YL YS
Hamlin	YP
Harrisville	YK

Wisconsin

Almena	YK
Antigo	JK
Appleton	JK JR
Aurora	JP
Baraboo	YP
Black River Falls	JL
Cameron	YJ
Clintonville	JL
Crandon	YL
Delafield	JS
Dodgeville	YR
Eau Claire	JP
Falun	JL
Fond du Lac	YR
Grantsburg	JL
Green Bay	JP YJ YP
Hager City	JL YL
Hancock	JS
Independence	YJ YS
Janesville	JR
La Crosse	JL YK YL YP
Lake Geneva	YK
Madison	JK YL YJ
Manitowoc	YL
Marshfield	YL
Medford	JS
Milwaukee	JK JL JP YJ YL YP YS QE QJ QR
Monroe	JL QP
Oshkosh	YJ
Platteville	YL YS QY
Plymouth	YK
Portage	YK
Prairie Farm	YS
Racine/Kenosha	JR

Reeseville	JL
Rhinelander	YS YR
Rice Lake	JR
Ripon	YS
Sand Creek	YR
Sheboygan	YJ
Sparta	JK
Tomah	JP
Two Rivers	JS
Verona	JP JS YS
Viroqua	JR
Waukesha	YR
Wausau	JP YK A5
W. Bend	JR
Westby	JS YR
Wisconsin Rapids	YJ

Wyoming

Baggs	YK
Casper	JL JP JR JS YJ YK YL YR A6
Cheyenne	JS
Cody	JR YK
Cokeville	JK JR
Evanston	JS YJ YR
Gillette	JK YK YL YP YS
Mountain View	JP
Newcastle	JK
Pinedale	JK YK YL
Rawlins	JK JL
Riverton	YP YS
Rock Springs	JR YP YS
Worland	JL JS

CANADA

Alberta

Alder Flats	XR XX
Algar Tower	JW XK XR YR
Amber	JK XS XY YP
Athbasca	XT YL
Banff	XP XY

Bear Canyon	JL XV XY
Beaverlodge	XR
Berland	XJ XP
Birch Mountain	JP XX YK
Blackfoot	JK JL JS XL XS
Bonanza	JW XR XU
Bonnyville	JP XK YR
Boyle	XJ YJ YS
Brazeau	XU YR
Brooks	JR XJ XP XT XW YJ YL YS
Calgary	JK JL JP JR JS JW XJ XK XP XR XU XX YJ YK YL YP YR YS
Calling Lake	XK
Camrose	JW XK XU YR
Canmore	JL JS
Cardston	YK
Caroline	JL JS XL XU
Cavendish	JR YJ
Cessford	JL JS XL XV
Chipewyan Lake	XP XS
Chipman	JK JL XL
Cochrane	JP YK
Cold Lake	JK XL XV YP
Coleman	JK XL XV
Conklin	JR XT YL
Consort	XV YL
Coronation	JP YK
Crossfield	XS XY
Debolt	JW XK YK
Drayton Valley	JR XJ XP XT XW YJ YL YS
Drumheller	JS
Edgerton	XK XU
Edmonton	JL JR JS JW XJ XL XP XR XS XT XV XW XX XY YJ YK YL YP YR YS QA QB QC QD QE QF QJ QK QO QP QR QY
Edson	JR XJ XP XT XW YJ YL YS
Elk Point	YJ YS
Exshaw	XT XW
Fairview	JR XW YS
Ft. Assinboine	JP XX
Ft. Chipewyan	JK JL

Ft. McMurray	JK JL JR JS XJ XT YJ YP
Ft. Vermillion	YL YS
Fox Creek	JP XR YK
Girouxville	XR YR
Gocan Lake	JR YS
Grande Cache	JP XK
Grand Prairie	JK JL JS XL XS XV XY YP
Granum	XJ YS
Hanna	JR XW
Hawk Hills	JP JW XK YK YR
High Prairie	JP
Hilda	JW XR
Hinton	JL JR JS XL XV YP
Hussar	JP JR XK XU YR
Indian Cabins	XW YJ
Innisfree	XV YP
James River	JP XK XU YK
Jenner	JP XK XU YR
Keg River	JK YP
Killam	XR XX
Kirby Lake	JW XR
Lac La Biche	YK XU XX
Lethbridge	JK JL JS YP
Little Buffalo	JR XJ XP YS
Little Smoky	JL JS
Lodgepole	JL JS XS XV XY YP
Lomond	JW XR YK
Lone Star	JL JS XL XS
Longview	XL XV
Manyberries	JR XT YJ
Marten Mountain	JK JL JS XL XS YP
May Tower	JL JS YP
Medicine Hat	JK JL JS XL XV YP
Moose Prairie	XW
Muskeg	JR YJ YL YS
Nanton	JP XU
New Dayton	JW XR YK
Nipsi	XR YR
Niton Junction	XR XX YK
Nordegg	JP XR XX YK
N. Habay	JP XK YR
Olds	XR YJ YL YR

Oyen	XX YK
Panny River	XR XT XV XY
Peace River	YJ YL
Pelican	JP XX YK
Pincher Creek	JR YJ YL
Provost	JR XP XW YL
Ralston	XS XY
Red Deer	JK JL JS XL XS XV XY YP
Red Earth	JP JW XK XU YK
Rimbey	JW YR
Robb	XS XY
Rocky Mountain House	JR XJ XP XT XW YJ YL YS
Rosebud	JK JL XL XV YP
Rycroft	JP XX
Schuler	XP YS
Smith	JW YR
Smoky Lake	YP
Snuff Mountain	XX
Steen River	YK
Stettler	XJ YJ YL YS
St. Paul	XJ XW
Sundre	JW XR XX YR
Swan Hills	JR XJ XP XT XW YJ YL YS
Taber	JP XK XU XX YR
Torrington	XP XT
Triangle	JW
Trochu	JP XK XU YR
Trout Mountain	JL JS XL XV
Two Creeks	XK XV YR
Valley View	XK XU YJ YL
Vegreville	JR XJ XP XT YL
Vermillion	JW XR XX YK
Viking	XW
Wabamun	XK XU
Wabasca	JW XK XR YR
Wainwright	XJ XT YJ YS
Walton Mountain	JP YR
Watt Mountain	JL JS XL XV
Weberville	XV YP
Wembley	XJ XP XT XW
Westlock	XR YK
Wetaskiwin	JP XR XX YK

Whitecourt	JK JL JS XL XS XV YP
Worsley	XT YJ YL
Youngstown	XL XU
Zama Lake	JR XJ XP XT YJ YL YS

British Columbia

Abbotsford	JS
Albert Canyon	JW
Alert Bay	JK JP YP
Bamfield	YS
Barriere	YL
Beatton Siding	JL
Beaverdell	YP
Beavermouth	YP
Bell Irving	JL
Bella Bella	JK
Blue River	YR
Bob Quinn	YJ
Britannia	JS
Brown Bear	YJ
Bull Moose	JP
Burnaby Island	JL
Burns Lake	JK JL
Cabin	YK
Cache Creek	JK
Calvert Island	JW
Campbell River	JJ JL JS YK
Caribou Hide	JW
Cassiar	YR
Castlegar	JL
Chase	YP
Chetwynd	JL YP
Chicken Neck	JS
Chilanko	JJ YL
Chilliwack	JK JS
Clearwater	JK
Clinton	YR
Coast Cone	YK
Compass Hill	JW
Courtenay	JP
Cranbrook	YJ
Creston	JP

Dawson Creek	YK YR
Dease Lake	YJ
Deer Ridge	YJ
Doig River	YS
Duncan	JS
Duncan Lake	JL
Earls Cove	YS
Elkford	YK
Elko	JL
Estevan Point	JP
Falkland	JJ
Faquier	YL
Fernie	JS
Fontas	JW
Ft. Nelson	JP JR XP XW YJ YL
Ft. St. James	JW YL
Ft. St. John	JK JS YJ
Ft. Ware	YP
Fraser	JK
Glacier Park	JK
Gleam	JK JS XY
Golden	YJ
Goldstream	YS
Goodlow	JP
Grandforks	YJ
Gwillin Lake	JS
Hagensborg	JL
Hazelton	JS
Hedley	YR
Holberg	YJ
Hope	YJ
Houston	JP YR
Hudson's Hope	YL
Invermere	JS
Ishkeenickh	JP
Iskut	YK
Kamloops	JR YJ
Kelowna	JW YK
Keremos	YP
Kingfisher	YP
Kitmat	YJ
Kitsault	JW

Klappan	JS
Lava Lake	JR
Lillooet	JW
Logan Lake	YK
Loos	YS
Louise	YS
Lumby	YL
Lytton	JS YR
Manning Park	JK
Mason River	YK
Masset	JR YJ
Mc Bride	YR
Mc Kay Range	JJ
Mc Kenzie	JP YR
Mc Leod Lake	JK
Meehaus	JP
Merritt	JL
Mica Creek	YJ
Minaker River	JR YJ YS
Mould Creek	JL
Mt. Dixon	JR
Muncho Lake	YJ
Nakusp	JS
Nanaimo	JK
Nazko	JK
Nechako	YR
Nelson	JS
New Denver	YP
New Westminster	JL
Nimpo Lake	JR
Nitinat	JW
Noble Mountain	JW
Nootka Island	YL
Oliver	JS
Omega Hill	YK
One Hundred Mile House	JS YP
Ootsa Lake	YK
Parksville	YP YR
Pemberton	YR
Penticton	JL
Pine Valley	JW
Port Alberni	YJ

Port Alice	JL JS
Port Hardy	JJ YL
Port Simpson	YK
Prespatou	JJ
Prince George	JJ JL JW YJ YK
Prince Rupert	JS YL
Preinceton	YS
Quesnel	JR YL YS
Red Rock	JP
Revelstoke	YJ
Riondel	YJ
Rivers Inlet	JP
Rossland	JK
Salmo	YR
Salmon Arm	JS
Sandspit	YK
Sarah Point	JK
Sayward	JR JW YJ
Sechelt Inlet	JW
Sidney	YK
Slocan	YR
Smithers	YJ
Sparwood	YR
Spence's Bridge	JS
Squamish	YS
Stewart	YP
Stuie	YS
Summit Lake	YS
Switt River	YJ
Terminus	YL
Terrace	JK JL
Thunder Mountain	JJ
Tofino	JS
Trutch Island	YR
Tsinhia	JK JS
Tumbler Ridge	YJ
Valemount	JK
Vancouver	JJ JP JR JW YJ YS
Vanderhoof	JS YP
Vernon	JK JP
Victoria	JL YR
Wells	YP

Whistler	YJ
Williams Lake	JP JW YK
Willis Lake	JL
Wonowon	JP JW XK XU YK YR
Woss Lake	YK
Woss Mountain	YS
Yoyo	JL XL

Manitoba

Anola	YP
Ashern	JL
Belair	JK
Benito	YR
Bissett	JW
Boissevain	JR
Brandon	YJ YK YS
Churchill	YJ
Cobham River	JL
Cowan	YJ
Cranberry Portage	JK JS
Cryl River	YL
Dauphin	JP JW
Easterville	YL
Elkhorn	YL YR
Falcon Lake	YK
Foxwarren	YP
Goose Lake	YL
Hadashville	YL
Haywood	YP
Hughes Lake	YJ
Jackhead	YJ YS
Kelsey	JK
Long Point	JL
Magnet	YR
Melita	JS
Middleboro	YS
Moose Lake	YK
Morden	JR
Neepawa	JR
Norway House	JK
Notigi	JL
Overflow	JK

Oxford House	JP
Pilot Mound	JL
Pointe du Bois	JR
Portage La Prairie	YL
Riverton	JP
Roblin	YK
Rosenfeld	JK
S. Indian Lake	YL
St. Laurent	JS
Snow Lake	YL
Steinbach	JS
Tan Creek	JR
The Pas	JP YJ
Thibaudeau	JL
Thompson	JP YJ
Wabowden	JR YS
Wee Lake	JP
Whiskey Jack	JL
William River	JS
Winnipeg	JJ JL JP JR JW YJ YK YR YS

New Brunswick

Bathurst	XJ YJ
Boiestown	JP
Campbellton	XS YP
Caraquet	JW
Edmundston	XX YK
Florenceville	XJ YS
Fredericton	XJ XW YL
Grand Falls	JK XY
Hampton	YK
Hoyt	JP
Jacquet River	YL
Kedgwick	JR
Minto	JS
Moncton	JS XS XU XY
Nackawic	YK
Newcastle	JK XS XV
Richibucto	JP
St. John	XV XY YP
St. Stephen	JR XT
Sussex	YR

Newfoundland

Baie Verte	JL
Bay L'Argent	JR
Birchy Lake	XP
Bonne Bay	YL
Cape Broyle	JL
Catalina	JL
Centreville	JR
Clarenville	XS YR
Codroy Pond	JR YL
Corner Brook	JK JP JS XR
Deer Lake	YR
Dunn's Brook	JL
Gambo	YL
Gander	JS XY
Grand Falls	JR YR
Gull Lake	JL
Hearts Content	JK XV
Hermitage	YL
Long Harbour	JS
Margaret	YJ
Marystown	YJ YP
Millertown	YL
Mt. Carmel	JR
Parker's Cove	JL
Placentia	YL
Port aux Basques	YL
Portland Creek	YR
Ramea	YR
Rantem	XR
Red Rocks	YR
Sheffield	JL YJ
Shoal Harbour	XP
St. Anthony	YL
St. John's	JJ JS XU XY YJ YL YP
Stephenville	JL YR
Twillingate	JJ
Victoria	JP

Northwest Territories

Angus	JP

Arrowhead	JP
Arctic Red River	JS
Chick Lake	YL
Dixon	JR
Ebbutt	JK JR
Enterprise	JL
Ft. Good Hope	YJ
Ft. Norman	YL
Ft. Providence	YJ
Ft. Resolution	YR
Ft. Simpson	JW YJ
Ft. Smith	JL
Grassy	YJ
Hay River	JS
Inuvik	JR
Little Chicago	JR
Morrisey	JL
Norman Wells	JL
Parsons	JP JS
Payne	JS
Pine Point	JK
Pointed Mountain	JR JW YS
Poplar	JS
Port Radium	JL YJ
Rae	JL
Red Knife	YR
Saline	JL
Snare	YJ
Taglu	JK
Travaillant	YR
Tuktoyaktuk	JL
Tungsten	YR
Wrigley	YJ
Yellowknife	YJ YP

Nova Scotia

Amherst	JK
Bridgewater	YK
Dartmouth	JK JL XL XP XT XU XY YJ
Digby	JL
Halifax	JK JL XL XP XT XU XY YJ
Kentville	JR YL

New Glasgow	YL YS
Port Hawkesbury	JK
Sackville	YP YR
Shelburne	YL
Sydney	YL YP
Truro	JK JW
Windsor	YS
Yarmouth	JK YP

Ontario

Barrie	YJ YP
Belleville	JP YJ
Brantford	YK
Brockville	YJ
Burlington	QC QR
Chatham	YJ
Clinton	YJ
Cornwall	YR
Dryden	YJ
Dundas	JR QC
Ft. Frances	YJ
Hamilton	JR JW QA QC QK QR YP
Hawkesbury	YP
Hespeler	JK XJ YK
Hull	JL JR JS JW YJ
Kenora	YJ
Kingston	YR
Kitchener	JK XJ YK
London	JR YJ
Markham	QK
Mt. Hope	QC QR
N. Bay	YJ
Omemee	JK
Orillia	YJ YP
Oro	YJ YP
Oshawa	JR QA QC
Ottawa	JL JR JS JW YJ
Owen Sound	YJ
Pembroke	JP
Peterborough	JK
Port Hope	YJ

Sarnia	YJ YL
Simcoe	YL
Smiths Falls	JP
St. Catharines	JK
Sudbury	JR YJ
Thunder Bay	JS YJ
Toronto	JJ JL JP JS XK XL XP XS XT XU XV WW XX YJ YK YL YR YS QA QB QC QD QF QJ QK QO QP QR QY A12
Windsor	JK

Prince Edward Island

Alberton	YK
Charlottetown	JL JR
Montague	YJ
Summerside	YS

Quebec

Bonaventure	YJ
Carleton	YP
Chandler	JW
Chicoutimi	YJ
Donnacona	JK
Granby	YJ
Montmagny	YP
Montreal	JK JL JP JR JS JW YJ YK YL YP YR YS
Quebec City	JR YJ YL YS
Rimouski	JR YS
Riviere du Loup	JS
Sherbrooke	YJ
St. Jos. de Beauce	JP
Trois Pistoles	YL
Trois Rivieres	YJ YR

Saskatchewan

Abbey	YR
Asquith	YJ
Beauval	JR
Bengough	JJ
Besnard Lake	JW

Biggar	YL
Blaine Lake	YK
Buffalo Narrows	JS
Canora	YS
Carlyle	YP
Carrot River	YS
Carswell River	YJ
Chaplin	YS
Cluff Lake	YL
Cudworth	YL
Cumberland House	JW
Davidson	JP
Debden	JP
Denare Beach	YR
Denzil	JW
Esterhazy	JS
Estevan	JP JW YR
Eston	YL
Foam Lake	YP
Ft. Qu'Appelle	YR
Fox Valley	JJ
Goodsoil	YK
Grenfell	YJ
Hanley	JS
Hatchet Lake	YS
Hudson Bay	JP
Humboldt	JR
Kane Lake	JR
Kindersley	JJ JL JS
La Loche	YS
La Ronge	JJ JK
Lafleche	JR
Lake Alma	JR
Lampman	YJ
Lashburn	JJ YP
Leader	YP
Lucky Lake	YK
Lumsden	YK
Maple Creek	YK
McKenzie Falls	YJ
Meadow Lake	YR
Melfort	JS

Melville	YL
Milestone	YP
Molanosa	YS
Moose Jaw	YJ YL
Moosomin	JL
Mossbank	JP
Narrow Lake	JR YK YL
Neilburg	YR
Nipawin	JL
N. Battleford	JK JL
Oxbow	JJ JR YS
Pelican Narrows	JP
Pinehouse	JP
Preeceville	JR
Prince Albert	JK YP
Raymore	JJ
Regina	JK JL YS
Rockglen	JW
Rosetown	JR
Sandy Bay	YS
Saskatoon	JL JW YP YR
Sled Lake	JL
Smeaton	YR
Southend	YK
St. Walburg	YL
Stoney Rapids	JP
Stoughton	JK
Swift Current	JL YJ
Tisdale	JJ
Turtle Lake	JS
Unity	JP
Uranium City	JL
Val Marie	JP
Wadena	YR
Waskeslu Lake	JJ
Watrous	JK
Watson	YK
Weyburn	JS YK
Wilkie	YS
Yorkton	JK

Yukon Territory

Beaver Creek	YJ
Carcross	JS
Carmacks	YJ
Dawson	JK JL JS
Destruction Bay	JL
Elsa	JL
Faro	JL
Fox	JL
Haines Jct.	YJ
Hyland	JR
Kusawa	YR
Le Berge	JR
Macmillan Pass	JW
Hyland	JR
Mickey	YL
Minto	JL
Murray	JS
Rancheria	JL
Rat Pass	JL YK YR
Ross Rivers	YJ
Salmon	JR
Shilsky	JR
Stewart Crossing	YJ
Tagish	JR
Watson	YJ
White Mountain	YR
Whitehorse	YJ YL YS

S S S S S S S S S S S S S S S S S

Wilderness/Remote Area Telephone Service

In some wilderness, mountain, rural, desert, or other remote areas, there is no easy or commercially feasable way of bringing landline telephone service to individuals and companies situated there. The usual problem is the expense and effort of running miles of telephone poles through difficult terrain to service a relative handful of subscribers. Wide rivers, mountains, dense forest, lack of roads, severe weather make the installation of telephone poles a difficult task.

This is when two-way radio is used to connect these subscribers with civilization. The radios summon emergency aid, but mostly they are used to place and receive telephone calls. Check out cellular bands, also the IMTS frequencies (see pages 41 and 42), as well as the RCC frequencies (see pages 111 and 112). Also, try the Basic Exchange Radio Service (BERS) frequencies. The remote area BERS subscribers operate in the band 816.0125 to 820.2375 MHz, with the central stations in the band 861.0125 to 865.2375 MHz.

In some wilderness areas of the USA, such as the National Forests of the western states, telephone service is reported to be provided by the Dept. of Agriculture's Forest Service. Some of these facilities have been reported as operating on 166.575, 166.585, 166.675, and 172.35 MHz. Further information is needed.

Throughout much of Alaska, and in parts of Canada, the problem is being beyond the range of VHF/UHF. As in the Australian outback, telephone service means two-way SSB communications on HF channels with base stations that can patch calls through to landline phones.

Thanks to HF/VHF/UHF communications, telephone calls can be made and received in rural, remote, wilderness and other isolated areas beyond landline phone services.

Note that in Puerto Rico, there are extensive communications with radio telephones at fixed locations in rural areas. These communications use most RCC, IMTS, and Aero Telephone frequencies (454.025 through 454.975 MHz).

Remote & Wilderness Area Telephone Calls
Other than on VHF/UHF Common Carriers.

Alaska (USB mode)

Base Location	Base Freq.	Remote Sites
Anchorage	3183 kHz	2253 kHz
	3183	3365
	3183	5137.5
Bethel	2604	2629
	2604	2773
	2604	5204.5
Cold Bay	3241	2691
Cordova	2312	2632
Fairbanks	3167.5	3354
	3167.5	5207.5
Juneau	2784	2694
	3241	3357
Ketchikan	2604	2256
	3180	2776

King Salmon	3164.5	2466
Kodiak	2781	2474
Kotzebue	2601	2463
Nome	2784	2471
	5370	5207.5
Unalaska	3238	3362
	5370	5134.5

Canada (USB mode)

Alberta	2621.5 kHz	2621.5 kHz
	5486.5	5486.5
British Columbia	2160	2030
	3270	3171
	3300	3224
	3359	3213.5
	4543.5	4573.5
	4820	4610
	4865	5405
	5248.5	5313.5
	5405	5810
	5486.5	5486.5
	6825	6790
	7953	7804
	9315	9462
	12080	12181
Manitoba	4837	4837
	5289.5	5289.5
Newfoundland	2621.5	2621.5
	5486.5	5486.5
Northwest Terr.	2621.5	2621.5
	3299.5	3299.5
	3310	3310
	4630.5	4630.5
	5281.5	5281.5
	5289.5	5289.5
	5411	5411
	5436.5	5436.5
Ontario	2621.5	2621.5
	4837	4837
	5186.6	4964.5
	5289.5	5289.5
	7401.5	7547.5

	9081.5	9081.5
	11651.5	11651.5
Quebec	2621.5	2621.5
	3299.5	3299.5
	4630.5	4630.5
	4837.0	4837.0
	5486.5	5486.5
Saskatchewan	2621.5	2621.5
	5289.5	5289.5
	5486.5	5486.5
Yukon	2567.5	2567.5
	5436.5	5436.5

(Undoubtedly, cellular telephone frequencies will be used to provide telephone service to some remote areas in the U.S. and Canada.)

b b b b b b b b b b b b b b b b b

Cordless Telephones

Cordless telephones proved to be of enormous interest to scanner owners when the devices went on the market in the 1970's. The phones could be easily intercepted by all manner of amateur and professional snoops, busybodies, yentas, private eyes, nosy neighbors, and just about everybody else! This kicked off the pastime of recreational eavesdropping.

Some states made laws against deliberate monitoring of cordless phones. Recently they have also been declared unlawful to monitor according to federal edict. But violations are hard to prove unless they are gross,

Most cordless telephones seem to claim operating ranges somewhere between 700 and 1,500 feet. Fact is, persons using scanners having a good outside antenna system can copy them at considerably greater distances. Most cordless telephone owners don't realize their private telephone calls can be so easily monitored by outsiders.

such as playing or distributing taped conversations.

The most popular cordless phones sold after October of 1984 have pedestal (base) units operating with FM at 10 channels on 46 MHz, with the handsets operating on paired frequencies in the 49 MHz band. To relieve congestion on those channels, in 1995 the FCC allocated an additional 15 base unit channels on 43 and 44 MHz, with 15 handset channels on 48 and 49 MHz. In the instance of the 15 added channels, base unit frequencies are not paired with specific handset frequencies, so manufacturers are free to mix and match at will.

The base units transmit both sides of a conversation and have a transmitting range substantially more than the handsets. People therefore usually monitor only the base frequencies. Most units are advertised as having a range of 1,500 feet between base to handset. That's more than a quarter of a mile. A scanner with a good outside antenna may be able to copy cordless phones from a few miles away, depending on conditions.

A high-performance antenna designed for serious long-range scanner eavesdropping the 43 to 49 MHz channels pulls in low-powered signals. This is the MAX-46-CORD. It's from CRB Research, P.O. Box 56, Commack, NY 11725. Phone (516) 543-9169.

One reason people monitor 49 MHz anyway is to eavesdrop on neighbors' FM wireless baby monitors. They operate between 49.82 and 49.90 MHz. People let them run all the time, so they broadcast all the sounds within range of their microphones, including various mom/pop activities, day and night.

Some 46/49 MHz cordless phones incorporate voice scramblers to thwart eavesdropping. These are simple analog scamblers, and easily descrambled. For instance, the Ramsey SS-70 is an inexpensive accessory that plugs into a scanner's external loudspeaker jack to permit clear reception of analog-scrambled transmissions. It isn't intended to work on comms that are digitally-scrambled. The SS-70 is sold factory wired, and also in kit form. It's from Ramsey Electronics Inc., 793 Canning Parkway, Victor, NY 14564. Phone: (716) 924-4560.

It's easy to get a digital readout of the tones from telephone touch pads. This provides information as to

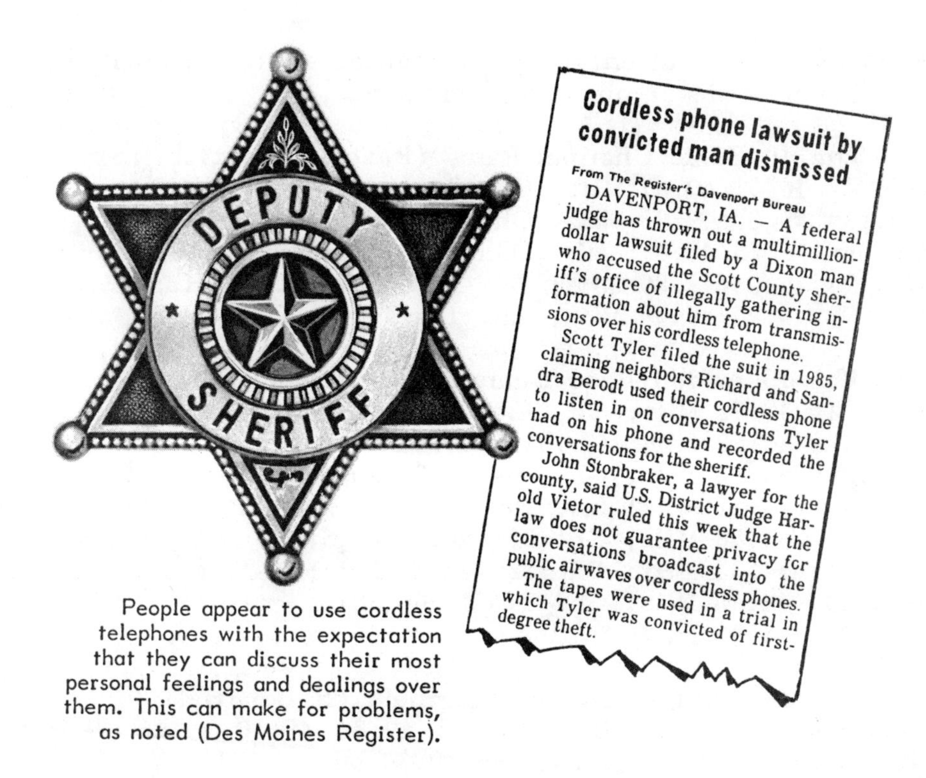

Cordless phone lawsuit by convicted man dismissed

From The Register's Davenport Bureau

DAVENPORT, IA. — A federal judge has thrown out a multimillion-dollar lawsuit filed by a Dixon man who accused the Scott County sheriff's office of illegally gathering information about him from transmissions over his cordless telephone.

Scott Tyler filed the suit in 1985, claiming neighbors Richard and Sandra Berodt used their cordless phone to listen in on conversations Tyler had on his phone and recorded the conversations for the sheriff.

John Stonbraker, a lawyer for the county, said U.S. District Judge Harold Vietor ruled this week that the law does not guarantee privacy for conversations broadcast into the public airwaves over cordless phones.

The tapes were used in a trial in which Tyler was convicted of first-degree theft.

People appear to use cordless telephones with the expectation that they can discuss their most personal feelings and dealings over them. This can make for problems, as noted (Des Moines Register).

numbers dialed, account numbers, and other data sent by buttons having been pressed on cordless phones. This comes via an outboard DTMF tone decoder such as the Optoelectronics DC440, or the MoTron TDD-8X. These connect by plugging into a scanner's recording jack. These are from Optoelectronics, 5821 NE 14th Ave., Fort Lauderdale, FL 33334. Phone (305) 771-2050; and MoTron Electronics, 310 Garfield St., Suite 4, Eugene, OR 97402. Phone: (503) 687-2118.

Deluxe "900 MHz" cordless phones all operate in the 902 to 928 MHz band, FM. Some use digital spread spectrum and other technologies not presently able to be copied on standard scanners. The FCC has not designated specific channels in this band, so various manufacturers have created their own systems.

Cordless phone users don't seem to realize that they can be so readily overheard by millions of scanner owners. Conversations monitored have brought criminal activity to the attention of law enforcement authorities, and been used as courtroom evidence.

The 10 Basic Channel Pairs (Base/Handset):
Ch. 1= 46.61/49.67 MHz; Ch. 2= 46.63/49.845 MHz;
Ch. 3= 46.67/49.86 MHz; Ch. 4= 46.71/49.77 MHz;
Ch. 5= 46.73/49.875 MHz; Ch. 6= 46.77/49.83 MHz;
Ch. 7= 46.83/49.89 MHz; Ch. 8= 46.87/49.93 MHz;
Ch. 9= 46.93/49.99 MHz; Ch. 10= 46.97.49.97 MHz.

The 15 Added Base Channels (Not Paired)
Ch. 11= 43.72 MHz; Ch. 12= 43.74 MHz;
Ch. 13= 43.82 MHz; Ch. 14= 43.84 MHz;
Ch. 15= 43.92 MHz; Ch. 16= 43.96 MHz;
Ch. 17= 44.12 MHz; Ch. 18= 44.16 MHz;
Ch. 19= 44.18 MHz; Ch. 20= 44.20 MHz;
Ch. 21= 44.32 MHz; Ch. 22= 44.36 MHz;
Ch. 23= 44.40 MHz; Ch. 24= 44.46 MHz;
Ch. 25= 44.48 MHz.

The 15 Added Handset Channels (Not paired)
48.76 48.84 48.86 48.92 49.02 49.08 49.10 49.16 49.20 49.24 49.28 49.36 49.40 49.46 49.50 MHz

Selected 900 MHz Phone Operating Frequencies
(Courtesy: Countermeasures newsletter)

Escort 9000/9010: Spread Spectrum unit noted on 24 frequencies between 905.1975 and 921.1985 MHz.

AT&T Model 9120: Base operates 902.00 to 905.00 MHz; handset 925.00 to 928.00 MHz.

Otron CP-1000: Base operates 902.10 to 903.90 MHz; handset 926.10 to 927.9 MHz.

Samsung SP-R912: Base: 903.00 MHz; handset 927.00 MHz.

V-Tech Tropez DX900: Base on 20 channels spaced at 100 kHz between 905.60 to 907.50 MHz; handheld on paired frequencies spaced at 100 kHz 925.60 to 927.40.

Panasonic KX-T9000: Base on 60 channels spaced at 30 kHz between 902.10 and 903.87 MHz; handset paired frequencies spaced at 30 kHz from 926.10 to 927.87 MHz.

? ? ? ? ? ? ? ? ? ? ? ? ? ? ? ? ?

Telephone Maintenance & Repair Services

Actual telephone calls aren't placed on these frequencies, they are authorized for use (in the United States) by telephone company personnel performing installations, service, repair, and maintenance to communications facilities.

If you want to know what Ma Bell and her relatives are up to, these are the frequencies to monitor-- especially if you've chosen to ignore the last three letters they sent you demanding payment of your phone bill. Why-- what's that telco van doing in front of your house?

Telephone Company Linemen & Maintenance Operations

35.16 43.16 151.985 158.34 451.175 451.225 451.275 451.30 451.325 451.35 451.375 451.40 451.425 451.45 451.50 451.525 451.575 451.625 451.675 462.475 462.525 MHz

Offset channels available on a restricted basis :

451.1625 451.1875 451.2125 451.2375 451.2625 451.2875 451.3125 451.3375 451.3625 451.3875 451.4125 451.4375 451.4625 451.4875 451.5125 451.5375 451.5625 451.5875 451.6125 451.6375 451.6625 451.6875 462.4625 462.4875 462.5125 (+ channels exactly 5 kHz higher, i.e. 456.1625, 456.1875 MHz, etc.).

UHF-T band channels available on a restricted basis in certain metropolitan areas (25 kHz channel spacing):

471.3125 to 461.4125, 472.9625 to 472.9875, 478.9625 to 478.9875, 507.3125 to 507.4125, + channels exactly 3 MHz higher, i.e. 47**4**.3125, 47**4**.3375, 47**4**.3625 MHz, etc.

Offset and UHF-T band channels, although available, are in little use at this time, and then only for low power handheld units. Most telephone maintenance communications take place on the primary channels (35.16, 43.16... listings).

8 8 8 8 8 8 8 8 8 8 8 8 8 8 8 8 8

1-Way Radio Paging Operations

One-way radio paging has been around for several decades, but the primitive little shirt-pocket gizmos that did nothing more than go "beep" have been upstaged by many new paging services that are far more sophisticated. These developments have caused the radio paging industry to expand at a very rapid rate during the past couple of years. Today, lots of people are wandering around or driving along with little black boxes in their pockets or on their belts-- physicians, real estate sales people, attorneys, farmers, electricians, appliance repair technicians, elected officials, executives, even drug dealers and law enforcement personnel-- in fact, anybody who feels that someone might need to contact them while they're not by a telephone.

It takes only seconds, and at relatively low cost, for a

Motorola is probably the world's most famous producer of radio paging receivers. This is their Bravo numeric display unit. The Bravo has a 12-digit LCD readout that provides the number of the person who called, plus other info like battery status, how many messages you have, and which one of the unit's several alerts has been selected for use.

Motorola's PMR-2000 is a personal message receiver that can deliver alphanumerics up to 32-charachers in length.

person to be alerted to the fact that someone-- a family member, co-worker, friend, customer, client, or boss-- wants to tell them or ask them something as soon as possible. Moms even give them to the kids to remind them that it's time to come home for supper. Or maybe Dad carries one so that Mom can remind him that the fishing trip's over and it's time to start up the outboard and head home. Possibly Mom's carrying a pager so Dad can let her know when it's time to put the tennis racquet away and come home to find his blue and red necktie for him. The uses of these devices are limitless, and for only a few dollars per month the service is available to anybody and everybody. And it looks like almost everybody decided it was a good idea!

The original "it only goes beep" pager, of course is still around, but the newer units can do soo much more. One unit simply vibrates to catch your attention rather than emitting a loud beep-- just right for use in church, an office, or restaurant. Some can display (via LCD's) the telephone number of the person attemnpting to reach you, or perhaps an entire message which can be held in the pager's memory for later recall-- it can store several telephone numbers and messages of from 80 to 160 characters each.

Some beepers have several alerting methods that can be switched into use-- loud/soft beep, hum, vibrate, and produce a distinctive sound in the event someone's trying to call with an urgent or emergency message.

There are beepers that remain silent until they're triggered into action by a call, and then the actual voice of the caller is heard giving the message. So many variations on the pager theme have been devised, that it's possible only to point out here some of the more popular features and types.

The control center for the transmission of such messages is the base station that sends out the paging signals, voice and/or a series of tones. The base station may be operated by a Common Carrier (that is, a telephone company, or wireline service), a Radio Common Carrier (an independent--non-wireline-- communications service offering its facilities to subscribers), a hospital, or a private business paging only its own employees.

This explosion in radio paging popularity has caused the frequency spectrum from 35 MHz upwards to be filled with the voices and cryptic tones of these services. In some areas there's even a frequency shortage for radio paging purposes. To meet this need, a few years ago the FCC allocated spectrum in 900 MHz portion of the spectrum for both non-commercial and private carrier paging systems (PCPS), with provisions for future interpool sharing.

Radio paging transmitters decicated to serving the general public are invariably located atop the tallest buildings and highest mountains in order to obtain maximum signal range. A major paging company in New York City offers contour maps showing its VHF alpha-numeric and digital display coverage extends 85 miles to the north, 55 miles to the west, and 120 miles to the south and the east. Their "UHF Extended Service" for tone, digital and alpha-numeric paging covers 75 miles to the east, 25 miles to the north, and 55 miles to the east and the west. Their "UHF Super Service" tone paging signals are claimed to reach 150 miles to the north and the south, 55 miles to the west, and 75 miles towards the east.

In addition to the radio paging frequencies listed here, there are also paging signals sent out via the Radio Common Carrier (RCC) service, listed elsewhere in this directory.

The clever folks who cooked up the ECPA law didn't much care if anybody wanted to sit there and listen to the tone-type paging signals, but they did make it a point to include in their law a little no-no relating to people listening to voice paging signals. Inasmuch as the two types of paging signals are often sent out on the same frequencies, it's hard to fathom the rationale of such an intended restriction-- even if the ECPA could be enforced-- even if there were any agency interested in enforcing the law.

One would have to assume, therefore, that those who broke their little pencil necks to pass this law surely must have come to the distinct conclusion that there are quite obviously many voice paging messages being sent out that are of such a highly personal or sensitive nature that they'd best not be overheard by third parties.

If that's what they think, then maybe it's true. Who am I to argue?

Department of Sneakyness: Some simple beepers are designed to display only numerals, with the idea being that those who want to send actial messages will upgrade to more sophisticated alpha-numeric pagers. Lots of folks with numeric beepers have figured out that they can send messages by creating their own codes. Instead of sending their telephone number, they may send 000-0177, which means, "skip your next scheduled call and go right to the next one;" or 000-0015, meaning "parts you were waiting for just arrived here," and so on. Not a bad idea at all.

Popular Radiopaging Channels (Voice/Non-Voice)

Medical & Emergency: 35.02 35.64 35.68 43.64 43.68 152.0075 157.45 163.25 453.025 453.075 453.125 453.175 MHz

Business & Private Systems: 26.995 27.045 27.095 27.145 27.195 27.255; 49.82 to 49.90; 152.48 154.57 154.60 154.625 157.74 158.46 462.75 462.7625 462.775 462.7875 462.80 462.8125 462.825 462.8375 462.85 462.8625 462.875 462.8875 462.90 462.9125 462.925 464.50 464.55 465.00 469.50 469.55 929.3625 929.3675 929.4125 929.4375 929.4625 929.6375 929.6625 929.6875 929.7125 929.7375 929.7625 929.7875 929.8125 929.8375 929.8675 929.8875 929.9125 929.9375 929.9625 929.9875; 935 to 940 MHz

Common Carriers & RCC's: 35.20 to 35.62; 35.66; 43.40 to 43.62; 43.66; 152.03 to 152.24; 152.51 to 152.84; 158.10; 158.70; 454.025 to 454.65; 930 to 932 MHz. (Links: 72.02 to 72.98; 75.42 to 75.98; 157.77 to 158.67; 459.025 to 459.65 MHz)

Note: Many non-voice paging systems operate in digital protocols such as POCSAG, SUPER POCSAG, or GOLAY. Alphanumeric beeper messages sent in these three formats can now be read using scanners. This is accomplished by using a decoder/reader like the Universal M-400, which connects easily by plugging into the scanner's recording output jack.

The Universal M-400 is a completely self-contained unit that reads out messages on a two-line 20 character LCD display, and also has a parallel printer port. The unit can read out DTMF codes, too. The M-400 sells for about $400. A decoder card that allows a PC to perform similarly is known as the Universal M-1200.

The non-voice beeper messages include not only call-back telephone numbers, but complete texts relating to business and personal matters. Some business messages may give you new insights into folks and occupations you knew only from news headlines. Meanwhile, many personal messages going out to alphanumeric beepers are either hilarious or simply bizarre.

For info on the Universal M-400 or M-1200, be sure to contact Universal Radio, Inc., 6830 Americana Pkwy., Reynoldsburg, OH 43068. Phone (614) 866-4267.

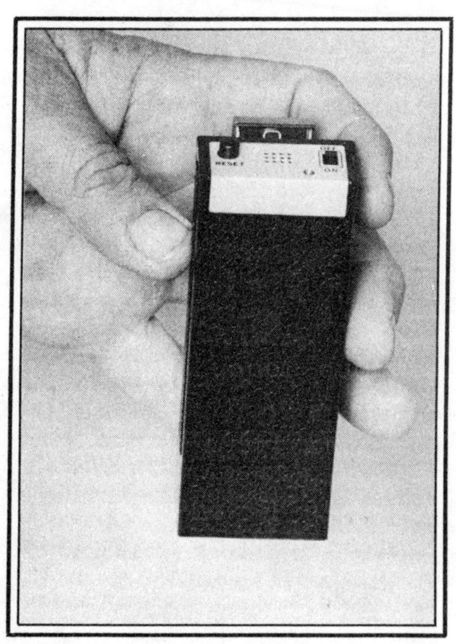

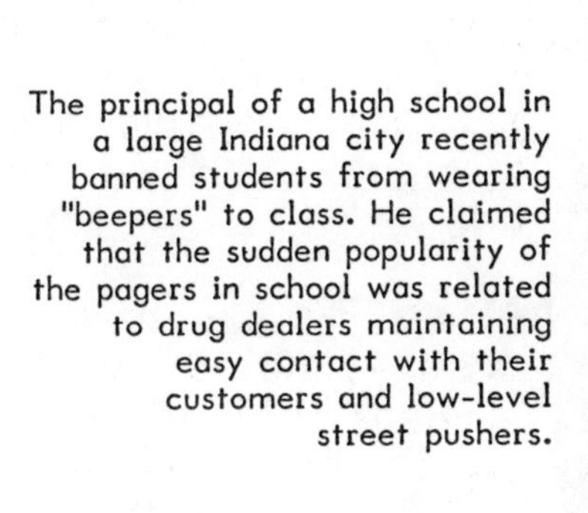

The principal of a high school in a large Indiana city recently banned students from wearing "beepers" to class. He claimed that the sudden popularity of the pagers in school was related to drug dealers maintaining easy contact with their customers and low-level street pushers.

Helpful Information

Monitoring publications of interest to scanner owners:

Popular Communications, *76 North Broadway, Hicksville, NY 11801. Phone: (516) 681-2922.*

Monitoring Times, *P.O. Box 98, Brasstown, NC 28902. Phone: (704) 837-9200.*

National Scanning Report, *Box 360, Wagontown, PA 12376. Phone: (610) 273-7823.*

Popular Electronics, *500-B Bi-County Blvd., Farmingdale, NY 11735. (516) 293-3000.*

Radio Monitors Newsletter of Maryland, *P.O. Box 394, Hampstead, MD 21074.*

9 9 9 9 9 9 9 9 9 9 9 9 9 9 9 9 9

Radio Common Carriers

Radio Common Carriers (RCC's) are FCC licensees theoretically able to provide a number of communications services including one-way paging with tones/voice, also two-way message exchange with the RCC operator (usually located at the offices of a telephone answering service). In many instances, subscribers to the RCC's services could also make and receive actual telephone calls in a manner similar to the telephone company's 152 MHz and 454 MHz IMTS mobile telephone operator services. However, RCC's are not telephone companies (wireline services), they are basically private operators (non-wireline) of radio communications services for hire which may or may not be directly interconnected to the telephone lines to the extent that their customers can send/receive telephone calls.

The advent of and fast expanding popularity of cellular service has appeared to cause a significant change in the status of RCC's, at least in areas where cellular service is available. Fact is that the cellular service looks to pretty much have left the RCC's in those areas to mostly drop out of the two-way message exchange and mobile telephone business and, instead, devote most of their communications efforts to providing one-way voice and/or tone coded radio paging.

The listing of RCC frequencies shown here may well be a sea of one-way radio paging stations in your own area, with nary a two-way exchange or mobile telephone call in evidence.

Radio Common Carrier Channels

Channel	Base Freq.	Mobile Freq.
1	152.03 MHz	158.49 MHz
3	152.06	158.82

5	152.09	158.55
7	152.12	158.58
9	152.15	158.61
11	152.18	158.64
13	152.21	158.67
21	454.025	459.025
22	454.05	459.05
23	454.075	459.075
24	454.10	459.10
25	454.125	459.125
26	454.15	459.15
27	454.175	459.175
28	454.20	459.20
29	454.225	459.225
30	454.25	459.25
31	454.275	459.275
32	454.30	459.30
33	454.325	459.325
34	454.35	459.35

Canadian RCC activity noted on: 152.24 163.47 163.74 164.355 164.37 164.43 168.54 MHz & others.

10 10 10 10 10 10 10 10 10 10 10 10

VHF-FM Local Marine Operator Telephone Calls

Private and commercial vessels on inland waterways, inland lakes, the Great Lakes, the Intracoastal Waterway, and coastal waters (as far out as 20 to 40 miles offshore) make considerable use of the facilities of VHF-FM Marine Operators operating on nine channels in the 161.00 to 162.00 MHz portion of the spectrum.

These operators dot the North American shoreline and the banks of larger rivers and lakes. While the listing here is believed to be relatively complete, it must be remembered

During the summer months, the primary users of these channels are the recreational boaters. The remainder of the year, there are lots of tugs, trawlers, and coastal tankers.

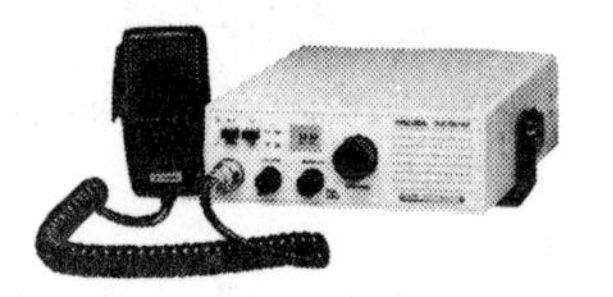

Priced at the low end, the Ray Jefferson #5000M is a 50-channel VHF-FM marine transceiver that can usually be found selling for about $150.

that new stations are constantly being placed in service and so you may well discover stations not listed herein.

The services of these VHF-FM band marine operators are less sophisticated than those to persons using car phones. For instance, when placing a call, the skipper must summon the operator by voice (and it may take the operator a while to reply), and then say aloud the number being called-- no automatic dialing here at all.

Most of these operators can send out ringer (selcall) tones, although commercial vessels are usually the only ones equipped to respond to such signals. Inasmuch as all vessels are required to monitor Channel 16 (156.80 MHZ) at all times, having selcall capabilities aboard means that a second VHF receiver must be in simultaneous operation-- one to maintain watch on Channel 16, with the other one set for selcall stand-by on the Marine Operator's channel.

In most cases, the Marine Operators simply announce the names of vessels for which traffic (shore/ship calls) are being held. This is done periodically on the operator's working frequencies, or the skippers expecting calls can check in with the operator and ask if there is any traffic. Only rarely will a Marine Operator attempt to call a vessel on Channel 16 (it's done only request of the calling party)-- however if it is done, once the initial contact is made, the actual call is handled on the operator's regular working frequency.

Calls handled through VHF-FM marine operators don't permit a conversational exchange of information. First, one

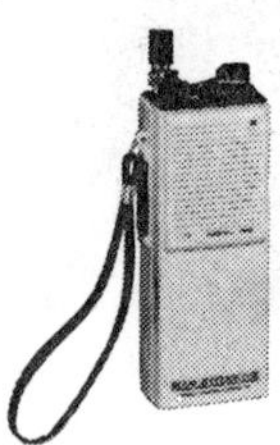

Ray Jefferson's model #789, a 78-channel VHF-FM hand-held usually selling for about $170. Puts out 3 watts.

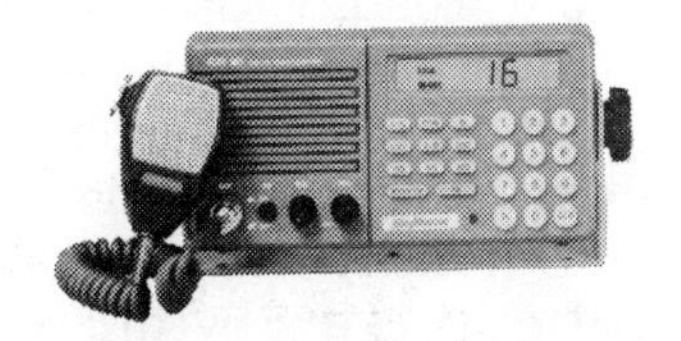

In the deluxe class of VHF-FM equipment is the Raytheon RAY-90. It has a list price around $1,000. Plenty of features, and 91 channels.

person speaks and then says "over," then the other person replies and says "over." Each time the person aboard the vessel wants to speak, a push-to-talk button on the microphone or handheld transceiver must be pressed.

Although a bit primitive, the whole thing does manage to be generally adequate, and since there isn't a monthly service charge for having a billing account with the Marine Operator (all you pay for is the calls you make, and not for any of the calls you receive), it's a bargain when compared to having a mobile telephone installed in your car--especially since you can buy a full-power 50-channel marine transceiver for around $150, or a 3-watt 78-channel handheld for about $170. Compare that with the price of buying a cellular telephone (although many new boats come equipped with optional cellular telephones in addition to VHF-FM transceivers).

Truth is, that although it's not only clearly illegal, but also rather sneaky, some folks have VHF-FM transceivers, or handhelds, that they use for making telephone calls from their vehicles! Once an account is opened with the Marine Operator's billing office, the operator has no way of knowing whether the caller is aboard a boat, or in a car, or wherever!

To open a billing account with a Marine Operator, a person would have to provide information as to the boat's registration number and FCC callsign. Those who own boats

The ICOM M-2 is a top-of-the line handheld. Runs 5 watts on 78 channels and sells in the $365 ballpark.

could readily provide such data, although total landlubbers might find it a hurdle to overcome.

What would ensue should the Marine Operator or the FCC ever catch someone trying something so underhanded isn't known since (even though there appear to be many who have gotten away with it for years), I've never head about anybody who has gotten caught. Undoubtedly the FCC would be decidedly unhappy on many levels and would have a half dozen rule violations to cite, followed by monetary forefeitures.

I suppose that, in trying a stunt so dastardly, one might restrict all use of the Marine Operator's facilities only to incoming calls; no billing account is required to receive calls, only to initiate calls. And, I'd guess that one would have to be somewhat creative in what was said over the air-- telling someone that you're stuck in heavy traffic at the corner of Main and 6th Streets, could possibly be regarded as suspicious to an alert Marine Operator, or to the many boat skippers listening in. However, when you hear some of the raunchy language that goes out during many marine telephone calls, you can't help but wonder if anybody really cares.

VHF-FM Marine Marine Operator Channels

Channel	Shore Freq.	Ship Freq.
24	161.80	157.20
25	161.85	157.25
26	161.90	157.30
27	161.95	157.35
28	162.00	157.40
84	161.825	157.225
85	161.875	157.275
86	161.925	157.325
87	161.975	157.375
88*	162.025	157.425

* Limited areas only: Puget Sound, Great Lakes (ex-Lake Michigan), & St. Lawrence Seaway.

Local Public Coastal/Inland Waterways Maritime (VHF-FM)

Alabama

Calvert	24 25
Coden	25 26
Demopolis	84
Grove Hill	28 86
Mobile	28 87
Muscle Shoals	26
Myrtlewood	25 28
Tuscaloosa	27

Alaska

Boswell Bay	26
Cape Spencer	26
Cold Bay	26
Craig	25
Diamond Ridge	26
Dillingham	26
Duncan Canal	27
Egegik	24
Juneau	26
Ketchikan	28
Kodiak	26
Lena Point	25
Manley	24
Metlakatia	86
Nikishka	28
Nome	26
Ratz Mountain	26
Seward	28
Sitka	28
Unalaska	28
Valdez	28
Yakutat	28

Arkansas

Blue Mountain	26
Blytheville	28
Helena	27 28
Little Rock	26
Watson	25

Maine

Maryland

Massachusetts

S. Yarmouth	28 84

Michigan

Bay City	28
Charlevoix	26
Copper Harbor	86 87
Detroit	26 28
Frankfort	28
Grand Marais	84 87
Harbor Beach	86 87
Hessel	84 86
Ludington	25
Marquette	28
Marysville	25
Monroe	25
Muskegon Heights	26
Ontonagon	84 86
Port Huron	25
Rogers City	26 28
Sault Ste. Marie	26
Spruce	84 87
St. Clair	84 86
Stevensville	85 86
St. Joseph	24
Tawas City	26

Minnesota

Duluth	84 87
Hasting	28
Minneapolis	26 28
St. Paul	26 28

Mississippi

Columbus	24
Greenville	26 84
Gulfport	28
Luka	86
Natchez	26 27
Pascagoula	27
Rosedale	24 86
Vicksburg	24 28

North Dakota

Garrison	25
Killdeer	84
Parshall	27

Ohio

Ashtabula	28
Cincinnati	28
Cleveland	86 87
Hamilton	85
Ironton	28
Lorain	26
Marietta	28
Mingo Jct.	28
Oregon	84 86
S. Amherst	26
Steubenville	28
Toledo	25

Oklahoma

Arkoma	28
Ft. Smith	28
Ketchum	27
Tulsa	26
Westport	28

Oregon

Astoria	24 26
Brookings	27
Coos Bay	25
Newport	28
N. Bend	25
Portland	26
Rainier	28

Pennsylvania

Erie	25
Freedom	26
N. Huntington	26 27
Philadelphia	26 85
Pittsburgh	26 27

Utah

Lake Powell	28
Navajo Mountain	26

Virginia

Hampton	25 26 27 84
Norfolk	25 26 27 84 85 87

Virgin Islands

St. Thomas	24 25 28 84 87

Washington

Bellingham	28 85
Camano Island	24
Freeland	87
Olympia	85
Port Angeles	25
Seattle	25 26
Tacoma	28
Tumwater	85

West Virginia

Charleston	27
Moundsville	26
Point Pleasant	24 26

Wisconsin

Lacrosse	26
Madison	28
Port Washington	85 87
Sturgeon Bay	86 87

Canadian VHF-FM Maritime

British Columbia

Alert Bay	26
Bull Harbour	26
Comox	26
Prince Rupert	26
Sandspit	26

11 11 11 11 11 11 11 11 11 11 11 11

Regional HF Coastal Telephone Stations

In the days before the VHF-FM marine band became widely implemented, most ship-to-shore telephone calls were handled in AM-mode on the 2 MHz band by Marine Operators located in virtually every major port and harbor. Those days are long gone, and there have been many changes on 2 MHz.

Those Marine Operators remaining active in this band have dropped AM mode in favor of more efficient USB mode. The band, in fact, does offer better range than VHF-FM and therefore it is primarily used for placing telephone calls by commercial vessels operating in inland waterways or coastal waters beyond the operating range of VHF equipment.

Daytime range on 2 MHz is less than 100 miles, but at night it is possible for ships to contact shore stations from more than 1000 miles at sea. Usually the shore station transmits both sides of conversations, and since these stations run far more power the ships (at night) their signals carry over great distances. And, just in case 2 MHz isn't suitable, some of these stations are capable of operation in higher frequency bands.

Regional Coastal Marine Telephone

	Shore (kHz)	Ships (kHz)
Alabama		
Mobile	2572	2430
Alaska		
Cold Bay	2312	2134
Cordova	2397	2237
Juneau	2400	2240
Ketchikan	2397	2237
Kodiak	2309	2131
Sitka	2312	2134
Florida		
Miami	2490	2031.5
Tampa	2550	2158
	2466	2009
Hawaii		
Kahuku	2530	2134
Indiana		
Jeffersonville	2086	2086
	2782	2782
	4116	4116
	6513	6513
	8725	8725
	13080	12233
	17299	16417
Louisiana		
Delcambre	2506	2458
	4366	4074
Massachusetts		
Boston	2506	2406
	2450	2366
	2566	2390
Michigan		
Rogers City	2514	2118
	2530	2158

	2582	2206
	4369	4077
	4381	4089
	4408	4116
	8794	8270
Missouri		
St. Louis	2086	2086
	2782	2782
	4408	4408
	6213	6213
	8737	8737
	13080	12233
	17299	16417
New York		
Buffalo	2514	2118
	2550	2158
	2582	2206
	4408	4116
	8794	8270
Ohio		
Lorain	4381	4089
	4408	4116
	8794	8270
Withamsville	2086	2086
	2782	2782
	4065	4065
	6513	6513
	8213	8213
	12333	12333
	16519	16519
Pennsylvania		
Pittsburgh	2086	2086
	2782	2782
	4065	4065
	6513	6513
	8213	8213
	12333	12333
	16519	16519

Puerto Rico		
San Juan	2530	2134
Tennessee		
Memphis	2086	2086
	2782	2782
	4089	4089
Texas		
Corpus Christi	2538	2142
Galveston	2530	2134
	2450	2366
Virgin Islands		
St. Thomas	2506	2009

Canadian Regional Maritime Telephone

British Columbia		
Alert Bay	2054	2054
	2458	2340
Bull Harbour	1630	1630
	2054	2054
Comox	1630	1630
	2054	2054
Prince Rupert	1630	1630
	2054	2054
	2060	2798
	2590	2166
Sandspit	1630	1630
	2054	2054
Tofino	1630	1630
	2054	2054
	2458	2340
Vancouver	1630	1630
	2054	2054

	2538	2015
	2558	2142
Victoria	1630	1630
	2054	2054
	2458	2340
Manitoba		
Churchill	2582	2206
	4375	4083
New Brunswick		
St. John	2582	2206
Newfoundland		
Cartwright	2582	2206
Comfort Cove	2538	2142
	2582	2206
Goose Bay	2582	2206
	4378	4056
St. Anthony	2514	2118
	2582	2206
St. John's	2514	2118
	2538	2142
	2582	2206
St. Lawrence	2514	2118
	2582	2206
Northwest Territories		
Cambridge Bay	2558	2142
	4363	4071
Coppermine	4363	4071
Coral Harbour	2582	2206
	4375	4083
Frobisher Bay	2514	2118
	2582	2206
	4375	4083
Inoucdjouac	2582	2206
Inuvik	2558	2142
	4363	4071
Killinek	2582	2206
	4375	4083
Resolute	2582	2206

	4375	4083
	8791	8267
Nova Scotia		
Canso	2514	2118
	2582	2206
Halifax	2530	2815
	2582	2006
Sydney	2530	2815
	2582	2206
Yarmouth	2538	2142
	2582	2206
Ontario		
Cardinal	2514	2118
Port Burwell	2514	2118
Sarnia	2514	2118
Sault Ste. Marie	2514	2118
Toronto	2514	2118
Thunder Bay	2514	2118
Prince Edward Island		
Charlottetown	2530	2815
	2582	2206
Quebec		
Grindstone	2514	2118
	2582	2206
Mont Joli	2514	2118
	2582	2206
Montreal	2514	2118
	2582	2206
Poste de la Baleine	2582	2206
Quebec City	2582	2206
Riviere au Renard	2514	2118
	2582	2206
Riviere du Loup	2514	2118
	2582	2206
Sept Iles	2582	2206

12 12 12 12 12 12 12 12 12 12 12

High Seas Telephone Service

High Seas telephone service is available from numerous stations throughout the world, although those listed here are the ones that normally handle calls for the United States and Canada. In addition to handling telephone calls to/from ships of all kinds (cruise ships, naval vessels, pleasure craft, tankers, freighters, research vessels, fishing boats, tugs, etc.) on the high seas, the High Seas stations also handle telephone calls to/from some oil drilling rigs and also to/from aircraft flying international routes.

High Seas stations use USB mode. They also use a full

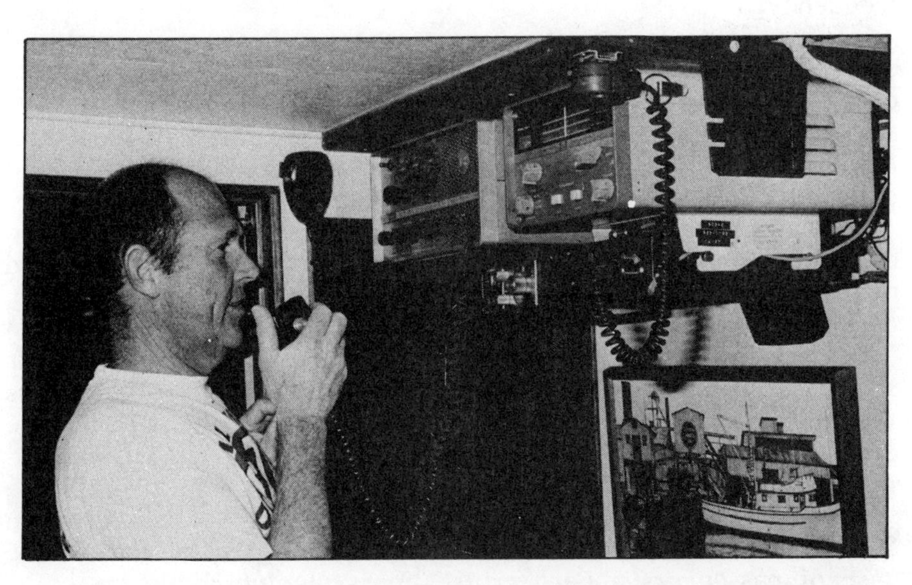

Passenger liners, tankers, cargo ships, ocean going yachts, and commercial fishing trawlers use high seas telephone service for placing telephone calls from mid-ocean or other distant offshore areas.

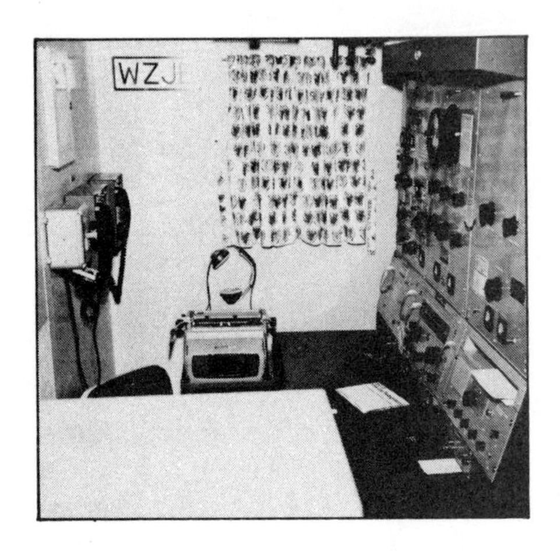

Here's the radio room of the M/V American Lancer, a 674 foot freighter. Its callsign is WZJB.

duplex (two channel) transmission system in which the shore station (usually) repeats the incoming signals from the received ship channel. Therefore, both sides of the conversation can normally be copied when monitoring only the shore station's frequency.

These stations continuously monitor all of their assigned channels for calls from vessels and aircraft, and they are also capable of sending out selective calling (selcall) tones to alert selcall-equipped vessels that they have calls for them. Moreover, these stations regularly announce the names and callsigns of those vessels for which traffic is being held.

As a general rule, frequencies in the 4 MHz band are used only at night, while 8 and 12 MHz frequencies are used day and night. Frequencies 16 MHz and higher would be used primarily during hours of daylight.

Interestingly, the fees the shore stations charge for their services are based upon a basic charge for handling calls, plus the toll charges for the landline call to the shore customer. The distance the ship is from the shore station isn't figured into the tariff at all. So, whether the ship working the California shore station is near the coast of Mexico, or in the Mediterranean Sea, there's no difference in the cost of the call. In fact, in the case of passengers aboard cruise liners placing High Seas calls, you'll hear the High Seas Operator tell the ship's radio operator ("sparks") the price of the completed telephone call.

Bands Used For High Seas Telephone (Worldwide)

Transmissions are USB mode:

Shore Stations	Ship Stations	Used Mostly
4351- 4438 kHz	4065- 4126 kHz	Nights
6501- 6525 kHz	6200- 6224 kHz	Nights
8707- 8815 kHz	8195- 8294 kHz	Nights
13077-13200 kHz	12230-12353 kHz	All hours
17242-17410 kHz	16360-16528 kHz	Days
22696-22855 kHz	22000-22159 kHz	Days
26145-26175 kHz	25070-25100 kHz	Days

High Seas Telephone Stations in North America

Mobile, Ala. (WLO)

Shore (kHz)	Ships (kHz)
4366	4074
4396	4104
4411	4119
8788	8264
8803	8279
8806	8282
13110	12263
13149	12302
13152	12305
17260	16378
17335	16453
17362	16480
22774	22078
22786	22090
22804	22108

Point Reyes, Calif. (KMI)

Shore (kHz)	Ships (kHz)
4357	4065
4402	4110
4405	4113
8728	8204
8743	8219
8782	8258
13077	12230
13080	12233
13083	12236
13161	12314

17245	16363
17248	16366
17311	16429
22735	22039
22762	22066
22777	22081
22801	22105

Fort Lauderdale, Fla. (WOM)

4363	4071
4390	4098
4405	4113
4423	4131
8722	8198
8731	8207
8746	8222
8758	8234
8791	8267
8809	8285
13092	12245
13098	12251
13101	12254
13110	12263
13143	12296
13164	12317
17242	16360
17266	16384
17269	16387
17272	16390
17287	16405
22738	22042
22741	22045
22759	22063

Manahawkin, N.J. (WOO)

4384	4092
4387	4095
4402	4110
4420	4128
8740	8216
8749	8225

8761	8237
8794	8270
13083	12236
13104	12257
13107	12260
13158	12311
17254	16372
17299	16417
17317	16435
17341	16459
22696	22000
22708	22012
22723	22027
22801	22105

St. Thomas, V.I. (WAH)

6510	6209
6513	6212
17245	16363
17248	16366
22762	22066

Vancouver, B.C., Canada (VAI)

4384	4092
4408	4116
6510	6209
6513	6212
8737	8213
8758	8234
13095	12248
17263	16381
22753	22057

Frobisher Bay, N.T., Canada (VFF)

4375	4083
6507	6206
8752	8228
13077	12230
17341	16459
22705	22009

Halifax, N.S., Canada (CFH)

4363	4071
4408	4116
6504	6203
6513	6212
8746	8222
13113	12266
13161	12314
17251	16369
17260	16378
22699	22003

High seas calls from merchant vessels sometimes get extremely personal as crew members advise wives of the lonliness of being at sea.

13 13 13 13 13 13 13 13 13 13 13

Oil Drilling Rig Telephone Calls

Oil drilling rigs can equal a small town in physical size and population. Still, it is by means of telephone calls sent by radio that those working on these rigs maintain contact with their offices and families.

Some oil rigs have been monitored passing their calls through the VHF marine band ship/shore phone ops. If close enough to the shore, they might even be able to pass their call via cellular phones.

The Offshore Radio Telecommunications Service was set up so that telecommunications companies on shore could establish central stations capable of operating on numerous frequencies. These frequencies are paired with other frequencies used by offshore subscribers located on oil drilling rigs, generally in the Gulf of Mexico.

This radio service operates with frequencies taken from locally unused UHF-TV Channels 15, 16, and 17, depending upon the area. Channel spacing is 25 kHz.

Generally speaking, operations in the area of Southern Texas utilize TV Channel 15. The shore station frequencies run from 476.025 to 477.975 MHz, with the

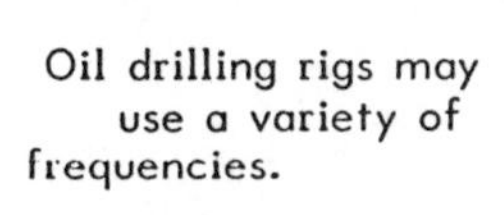

Oil drilling rigs may use a variety of frequencies.

offshore subscriber stations operating between 479.025 to 480.975 MHz.

UHF-TV Channel 16 frequencies are used in the Southern Louisiana/Texas area. The shore station operations are in the band 485.025 to 486.975 MHz, with the offshore subscribers using the band 482.025 to 483.975 MHz.

Southern Louisiana is where UHF-TV frequencies have been set aside for these operations. Shore stations are in the 488.025 to 490.000 MHz band. Offshore subscriber stations are on paired frequencies in the 491.025 to 493.000 MHz band.

Offshore stations in this service can't use more than 25 watts (ERP) if they are within 23 miles of the shore. Beyond that distance, they may use up to 100 watts. Offshore rigs are limited to antennas no more than 200 ft. above mean sea level.

14 14 14 14 14 14 14 14 14 14 14

Railroad Telephone Calls

Telephone calls from passenger trains on certain routes are now available and are handled through the standard cellular phone services. This service, under the trade name of Railfone, is now available between Washington, DC and Chicago, IL; between Washington, DC and New York City; and between Los Angeles and San Diego, CA. This service, aboard Amtrak trains, is provided by GTE Railfone Incorporated.

Some larger railroads have PBX (Private Branch EXchange) telephone service. This isn't a service available to passengers, rather it's primarily for the internal system use of railroad supervisory personnel. PBX enables them to place business calls from their vehicles, or receive them in their vehicles. Rather than the calls being handled through the communications facilities of Common Carriers, everything takes place on channels in the Railroad Radio Service, and through the private telephone switchboard of the railroads themselves-- which patch the calls through to outside telephones.

PBX's are two-channel systems, and the listings here show the repeater output frequencies of some of the systems

believed to be used by major railroads. PBX service on specific railroads may be available only on certain sections of their respective routes. All PBX input/output channels are located within the frequency band 160.215 to 161.565 MHz (in Canada from 159.81 to 161.565 MHz).

Railroad PBX Telephones

Selected Major Railroads:

AT&SF: 160.245 160.26 160.375 160.425 MHz
Burlington Northern: 160.425 160.62 161.13 MHz
Conrail: 161.13 161.445 MHz
CP Rail: 160.175 160.265 160.845 161.175 161.265 161.505
L&N: 160.98 161.34 MHz
Missouri Pacific: 160.605 160.755 160.815 MHz
N&W: 160.515 161.275 MHz
Richmond, Fredericksburg & Potomac: 161.22 MHz
Seaboard: 160.215 161.265
Southern Pacific: 160.35 160.68 160.80 160.95 160.89 161.22 161.34 MHz
Union Pacific: 160.29 161.28 MHz

15 15 15 15 15 15 15 15 15 15 15 15

Air/Ground Telephones

In addition to the telco air/ground phone services on 454/459 MHz, there are now more sophisticated services operating in other bands and used primarily by the commercial airlines. Most notably, these services are provided by GTE Airfone and by In-Flight Phone Corp., although the exact technologies used by the the several companies differ from one another to some considerable extent.

The general frequency bands used are 849 to 851 MHz for the ground station uplinks, and 894 to 896 MHz for the aircraft downlinks, with channels spaced in 6 kHz steps. In the ground-to-air uplink band, the lowest frequency voice channel begins at 849.0055 MHz, its paired air-to-ground channel is 894.0055 MHz. From these points, mark off channels every 6 kHz. There are some control channels within the series that are used for the ground and air units to exchange system operational data for the phones.

These systems use standard AM mode. The best way to monitor them is to set up the scanner to search (AM

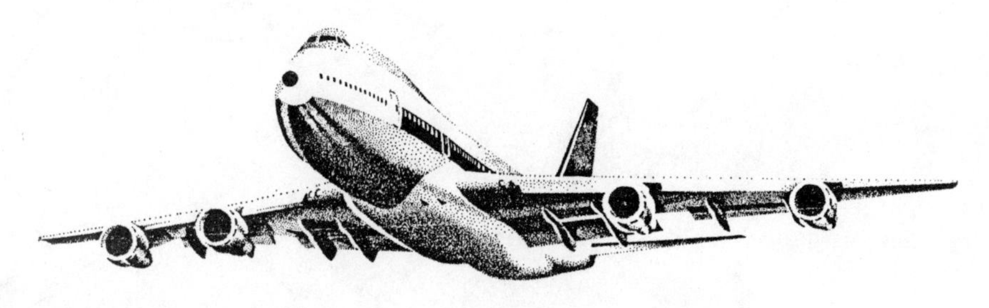

mode) in 5 kHz steps. You can try the uplink band, but may not hear any signals unless you happen to be in the vicinity of a ground station. Best bet is to search the 894.00 to 896.00 MHz downlink band, which lets you monitor the airliners. Flying at high altitudes, they can be heard over wide areas. Of course, you'll hear only one side of conversations.

For best reception, an outside UHF antenna is recommended, plus a signal preamplifier.

16 16 16 16 16 16 16 16 16 16 16

Military Aircraft VIP Telephone Calls

Conventional military aircraft from the MAC and SAC branches of the U.S. Air Force, and also U.S. Navy aircraft can frequently be monitored in communication with USAF ground stations asking for Autovon patches. Autovon is the U.S. military's private internal (primarily) landline telephone system.

Most of the Autovon patches to/from military aircraft relate to the exchanging of essential flight data-- weather, arrival times, fuel status, mechanical problems ("write ups").

However, there is another category of military air/ground telephone call that goes far beyond Autovon patches concerning flight data. These are calls to/from VIP aircraft, such as Air Force 1, Air Force 2, and various Special Air Mission ("SAM") flights.

The majority of such calls are placed between these aircraft and the ground station at Andrews Air Force Base ("Andy") in MD. Communications below 30 MHz are in SSB, with both USB and LSB modes used at various times. Those in the VHF and UHF bands utilize FM.

The UHF (400 MHz band) communications are apparently handled via a large and widespread network of ground station transmit/receive sites, all remotely controlled from Washington, DC. Equipment located in relatively close proximity to one of the 407.85 MHz ground transmitters would be able to pick up the ground station's half of the duplex communications exchange. In most cases, however, only the aircraft half of the contact (on 417.50 MHz) is heard-- and since these jets fly at high altitudes, the transmissions can be copied on the ground while the aircraft is a more than 200 miles distant. Of course, the aircraft are where the VIP's are-- the President, the Vice President, Secretary of State, and various governmental officials and high ranking military officers.

Many telephone calls are placed to/from regular (non-Autovon) telephone numbers, or to "Crown" (the White House Communications Center). Most calls are "in the clear" (that is, <u>un</u>scrambled). Conversations can cover a large number of topics from media strategy, to status updates on matters of national importance, to requests for specific personnel to be on hand when the aircraft lands, etc.

Calls on the shortwave bands take place while these aircraft are enroute to or returning from overseas points. Actually, the frequencies used are more plentiful than listed here, however these are ones recently monitored and will give you some general indications of the portions of the spectrum favored.

While the ID's Air Force 1 and Air Force 2 are well known, it should be noted that the following other ID's may also be noted placing telephone calls: SAM-01 = A VIP flight carrying a foreign head of state; SAM-26000 = Presidential backup aircraft without President aboard; SAM-27000 = Presidential aircraft, President not aboard; SAM-21682 = Vice President's aircraft, VP not aboard; SAM-86791 = Secretary of State aircraft; SAM-86972 = National Security Advisor's aircraft. Other SAM ID's noted frequetly include SAM-12492, SAM-31683, SAM-60200, and SAM-60202, among others.

Air Force 1 & 2, + Other VIP Military Aircraft Patches

Domestic flights (FM):
"Echo Foxtrot" Ground 407.85 MHz; Aircraft 415.70 MHz
"Yankee Zulu" Ground 162.6875; Aircraft 171.2875 MHz

Overseas flights (USB or LSB), ±3 kHz:

3116 kHz	6817 kHz	11055 kHz	13241 kHz
6715	6927	11210	13752
6730	9120	11239	15048
6756	11035	11249	16090
6761	11180	13215	18027

17 17 17 17 17 17 17 17 17 17 17 17

Military Affiliate Radio Service (MARS) Calls

MARS stations belong to networks sponsored by the various military services-- Army, Navy/Marines, and Air Force. Stations are usually staffed by persons holding Amateur Radio licenses. Although MARS stations may be located aboard vessels of the Navy and Coast Guard, and at American military installations in the United States and overseas, there are hundreds of MARS stations in the homes of ham operators who belong to either the Navy, Army, or Air Force MARS groups.

MARS frequencies are usually located just outside the edges of the ham bands, and they most often used for the exchange of personal messages and telegrams between American military personnel and their families at home. These messages are sent by voice, CW, RTTY, and even packet modes.

One of the ways MARS operates is by running telephone patches (usually USB mode) between ships at sea or overseas bases and stateside operators, who feed the incoming

MARS, the Military Affiliate Radio System, has divisions in each branch of the armed forces. Phone patches between service service personnel overseas and home are part of what takes place on MARS frequencies.

transmissions through the telephone lines to the families being called, and vice versa.

These calls are invariably rich in human drama-- personal and financial problems, births, deaths, divorces, etc. A frequency found active with such phone patches can go for several hours without a break, as military personnel line up to place a free long distance call to home.

MARS networks are especially busy at holiday times, or at any time there is a military crisis overseas.

The Canadian Forces Amateur Radio (CFAR) service is much the same in purpose and activity as MARS, and these networks are also widely reported.

Military (MARS) Telephone Patches

Most popular USAF MARS patch frequencies:

7633.5 kHz	14389.0 kHz	14877.0 kHz	20807.0 kHz
10267.0	14390.5	15632.0	20991.0
10270.0	14402.0	16452.0	23862.0
11407.0	14530.0	17670.0	27736.0
13614.0	14606.0	19226.0	27829.0
13927.0	14829.0	20188.5	27978.0
	14832.0		27991.0

Most Popular USN MARS patch frequencies:

13530.0 kHz	13975.5 kHz	14468.5 kHz	14478.5 kHz
13827.5	14443.0	14471.5	14760

Most popular US Army MARS patch frequencies:

13997.5 kHz
14403.5
14485.5

Most popular Canadian Forces Amateur Radio (CFAR) patch frequencies:

6905 kHz	14838.5 kHz
13971	20957
14384.5	20972

18 18 18 18 18 18 18 18 18 18 18

Amateur Radio HF Phone Patches

Amateur radio operators using the HF bands, where Autopatch service doesn't exist, can still handle phone patches from distant stations, including from hams aboard ships in international waters. These services are thanks to a manually operated phone patch, an inexpensive device that can turn any communications station into a point for the passage of telephone calls-- license authorization permitting.

Even CB stations (26.965 to 27.405 MHz band) are permitted to use phone patches, and have also been heard doing so in both AM and LSB modes.

Amateur Radio Phone Patch Telephone Calls

1840 to 2000 kHz band (LSB mode)
3750 to 4000 kHz band (LSB mode)
7150 to 7300 kHz band (USB mode)
14230 to 14350 kHz band (USB mode)
21200 to 21450 kHz band (USB mode)
24930 to 24990 kHz band (USB mode)
28300 to 29700 kHz band (USB mode)
50.10 to 51.00 MHz band (USB mode)

Quick Reference Recap Chart Of
Most Popular Scanner Frequency Clusters
(Lower/Upper Freq. Limits Given)
Frequencies Shown in MHz.

Lower		Upper	Use
35.22	-	35.66	Radio Paging
43.22	-	43.64	Radio Paging
46.61	-	46.97	Cordless Telephones
49.67	-	49.97	Wireless Room/Baby Monitors
145.10	-	145.50	Ham autopatch
146.61	-	147.39	Ham autopatch
152.03	-	152.21	RCC & Radio Paging
152.48	-	152.825	IMTS Telephone/Canada
152.51	-	152.81	IMTS Telephone & Radio Paging
157.74	-	158.10	Radio Paging
158.49	-	158.70	Radio Paging
158.91	-	161.565	Railroads/Canada
160.215	-	161.565	Railroads/USA
161.80	-	162.025	Marine Operator Ship/Shore
223.85	-	224.98	Ham autopatch
442.00	-	445.00	Ham autopatch
447.00	-	450.00	Ham autopatch
451.175	-	451.6875	Telephone Repair
454.025	-	454.35	RCC & Radio Paging
454.40	-	454.65	IMTS Telephone & Radio Paging
454.70	-	454.975	Telco air/ground operators
459.025	-	459.65	Radio Paging
459.70	-	459.975	Telco air/ground aircraft
479.025	-	493.00	Offshore Oil Platforms
861.0125	-	865.2375	Basic Exchange Radio Service
869.00	-	894.00	Cellular Phones
902.00	-	928.00	900 Mhz Cordless Phones
929.00	-	932.00	Radio Paging

19 19 19 19 19 19 19 19 19 19 19 19

Amateur Radio Autopatch Service

There are many benefits to having a ham radio license in the United States or Canada. One of the benefits the general public seldom hears about is something known as autopatch.

Autopatch, which is primarily available in the ham bands above 144 MHz, enables ham operators, via their own personal handheld or mobile transceivers, to place telephone calls. These calls are usually local (non-toll), and are (by FCC/DOC mandate) not permitted to be of a business nature.

Access to the telephone system is achieved through the facilities of certain private or club-owned repeaters what are equipped for autopatch operation. Upon properly accessing such a repeater, the ham can place the telephone call by pressing the push buttons on a keypad which generates the same dialing

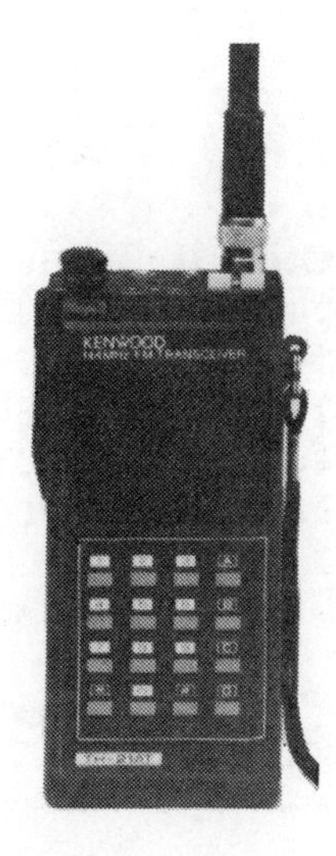

Ham radio operators have the ability to make telephone calls via the Autopatch facilities on many VHF/UHF repeater systems. This Kenwood TH-41AT handheld transceiver for the band at 440 MHz has a built in keypad that will generate the proper access and dialing tones for getting the calls through.

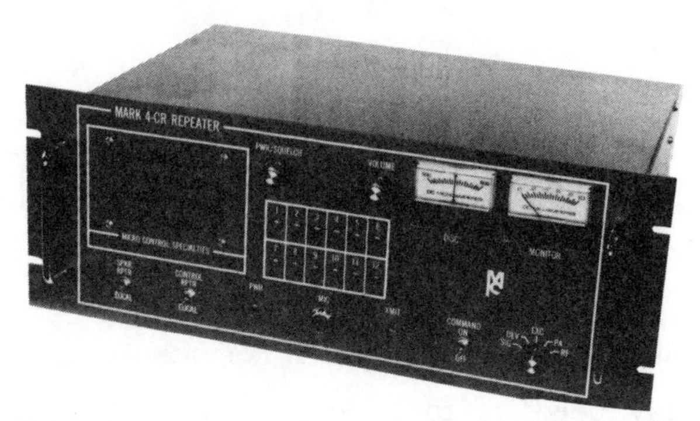

Mixed in with the general ham chatter on VHF/UHF bands are telephone calls placed through repeaters like this one.

tones as produced when placing a call from a standard home pushbutton telephone.

Of course, access to specific repeaters to make such calls is at the discretion of the owners of those facilities. Some repeaters are "closed," and can be accessed only by those transceivers that have been equipped to cause a closed repeater to respond to certain transmitted CTCSS (PL) tones.

The most complete listing of ham repeaters having autopatch facilities is the **ARRL Repeater Directory,** published by the American Radio Relay League, Newington, CT 06111.

Amateur Radio Autopatch

Output Freqs.	Input Freqs.
145.10-145.50 MHz	144.60-144.90 MHz
146.61-147.39	146.01-146.37
223.85-224.98	222.25-222.38
442.00-445.00	447.00-450.00
447.00-450.00	442.00-445.00

20 20 20 20 20 20 20 20 20 20 20

Satellite Telephone Calls

Modern technological methods have shifted a considerable amount of high seas telephone traffic, and virtually all transoceanic point-to-point telephone traffic off of the HF (4 to 30 MHz) bands and onto numerous satellites such as **Spacenet, Marecs A (Marisat), Satcom, Westar,** and others.

The frequency spectrum and transmission/modulation techniques involved are far beyond the design parameters of all present day scanners and communications receivers the way people use such equipment for monitoring HF/VHF/UHF communications. It is, however, feasable to rig up such equipment in connection with a TVRO (home satellite TV receiving station) and monitor this type of telephone traffic.

Of course, you'll need to have TVRO equipment, including a parabolic dish antenna-- and also the extensive information on hooking everything together and getting it going.

This highly specialized type of monitoring does have a

Many long distance telephone calls go through satellites. The public doesn't realize that, somewhere along the line, a large number of telephone calls are sent out via radio signals.

growing legion of enthusiasts, not the least of which are several intelligence gathering agencies and also apparent diplomatic personnel stationed here from several nations. But there are plenty of others listening there, too, including corporations and casual listeners such as hobbyists.

Satellite terminals for High Seas service run about $30,000 per ship, that's why so many vessels still use the HF bands. But, Holland America Lines, Cunard, Norwegian Caribbean, Royal Viking and several other major cruise lines are installing passenger telephones that work through satellites. These calls cost the passengers between $11.00 and $15.50 per minute.

The way to get started in monitoring telephone calls (foreign, domestic, and high seas) carried by satellite is to acquire the specialized equipment and information necessary. The standard (and best) reference sources for doing this are the following books: **The Hidden Signals on Satellite TV,** by Tom Harrington and Bob Cooper (Universal Electronics); **World Satellite Almanac,** by Mark Long (CommTek Publishing Co.); and **Communications Satellites,** by Larry Van Horn (Grove Enterprises).

Satellite Telephone Call Frequencies

High Seas Ship/Shore Calls:

Downlink Band	Uplink Band
1.537 - 1.543 GHz	1.638 - 1.645 GHz
4.194 - 4.201 GHz	6.420 - 6.425 GHz

Point-to-Point Overseas/Domestic Long Distance Calls:

3.700 - 4.200 GHz	5.925 - 6.405 GHz
11.700 - 12.200 GHz	14.000 - 14.500 GHz

21 21 21 21 21 21 21 21 21 21 21 21

Long Distance Microwave Relays

Much long distance telephone traffic today is carried over satellites or through fiber optic cables. Yet, a considerable amount of telephone call traffic still being carried over terrestrial microwave relays operated by AT&T, MCI, Western Union, and other Common Carriers.

An enormous amount of frequency space is dedicated to these operations, although the frequencies involved would require receivers and antennas of sophisticated design and installation. The signals involved travel in relatively narrow beams and, in any case, would be able to be picked by using equipment placed directly within the path of the signal beam.

Nevertheless, apparently telephone traffic going out over such facilities does appear to be regularly and diligently monitored by various intelligence agencies, as well as diplomatic legations of certain nations, and those involved in industrial espionage. The microwave signals in the areas of New York City, Washington, and San Francisco seem to be of maximum interest to those monitoring them.

Long Distance Telephone Microwave Relays

2110 to 2180 MHz band
3700 to 4200 MHz band
5925 to 6425 MHz band
10550 to 11700 MHz band
17700 to 23600 MHz band
31000 to 31200 MHz band

SECRET

SPECIAL ACTION REQUEST

Dear Reader,

Hope you found this book useful and interesting. I also hope that you'll call to my attention any additions, changes, and (not that you'll find many) errors. Your comments, philosophical observations, monitoring anecdotes, and ideas for future updated editions are also welcomed. Just drop me a card or letter in care of the publisher of this book and you'll be mentioned in dispatches.

Sincerely,

Tom Kneitel

SECRET
WARNING NOTICE —
INTELLIGENCE SOURCES AND
METHODS INVOLVED